INTELLIGIBLE
DESIGN

A Realistic Approach to the
Philosophy and History of Science

INTELLIGIBLE DESIGN

A Realistic Approach to the Philosophy and History of Science

Julio A Gonzalo

Universidad San Pablo-CEU de Madrid, Spain &
Universidad Autonoma de Madrid, Spain

Manuel M Carreira

Universidad Pontificia de Comillas, Spain

 **World Scientific**

NEW JERSEY · LONDON · SINGAPORE · BEIJING · SHANGHAI · HONG KONG · TAIPEI · CHENNAI

Published by

World Scientific Publishing Co. Pte. Ltd.

5 Toh Tuck Link, Singapore 596224

USA office: 27 Warren Street, Suite 401-402, Hackensack, NJ 07601

UK office: 57 Shelton Street, Covent Garden, London WC2H 9HE

British Library Cataloguing-in-Publication Data
A catalogue record for this book is available from the British Library.

INTELLIGIBLE DESIGN
A Realistic Approach to the Philosophy and History of Science

ISBN 978-981-4447-60-7

Typeset by Stallion Press
Email: enquiries@stallionpress.com

Printed in Singapore

Contents

Contributors/Editors

Manuel Alfonseca is a doctor in Electronics Engineering (1972) and Computer Scientist (1976), both degrees obtained at the Universidad Politécnica of Madrid. He is a professor at the Department of Computer Science of the Universidad Autónoma of Madrid, where he was director of the Escuela Politécnica Superior (2001–2004). Previously, he was Senior Technical Staff Member at the IBM Madrid Scientific Center, where he worked from 1972 to 1994. He has published over 200 papers and over 40 books on computer languages, simulation, complex systems, graphics, artificial intelligence, object-orientation and theoretical computer science, as well as popular science and juvenile literature, with different awards in all of these fields.

John Beaumont is a lawyer by training and was formerly Head of the School of Law at Leeds Metropolitan University, England. He is now working as a legal consultant and freelance writer on Catholic issues. He has written for leading Catholic journals in both the United States and Great Britain. His latest book is entitled *Roads to Rome: A Guide to Notable Converts from Britain and Ireland from the Reformation to the Present Day* (2010). It is published by St. Augustine's Press (www.staugustine.net). He is about to publish a book on American converts to the Catholic faith. John Beaumont can be contacted at john.beaumont7@virgin.net.

Father Manuel M. Carreira, SJ, after his degrees in Philosophy and Theology, obtained a Master in Physics (John Carroll University, Cleveland) and PhD in Physics (The Catholic University of America, Washington) with a thesis on cosmic rays directed by Dr. Clyde Cowan (co-discoverer of the neutrino with F. Reines, Nobel 1955).

He taught Philosophy of Nature at Comillas University in Madrid. In 1991, on the centenary of the Vatican Observatory, he gave a course in Rome on galaxies for 20 bishops from different countries. In 1986, John Carroll University awarded him its Centennial Commemorative Medal in recognition of his work at that center. In 1999, he received the *Medalla Castelao* of the Galician regional government, for promoting the prestige to the region with his cultural activities

His book *Filosofía de la Naturaleza, Metafísica de la Materia*, is used as a text in several educational centers in Spain and America. Two monographs, *El Creyente ante la Ciencia* (BAC) y *El Hombre en el Cosmos* (Sal Terrae) have been published in Spain.

Dr. Thomas B. Fowler works in the area of system analysis and design, including system stability and performance. He has spent many years doing objective research and analysis on behalf of the U.S. Government. He does research on network performance issues, chaos and fractals, telecommunications technology trends and the mutual interaction of these trends and changes in society as a whole, and is also actively involved in planning the next generation of U.S. government telecommunications for clients such as the General Accounting Office, the U.S. Senate, and the General Services Administration. He also serves as an adjunct professor in the Electrical and Computer Engineering school at George Mason University, where he teaches graduate courses in optics and optical communications. He has a doctorate in systems and control theory from the George Washington University, a Master of Science from Columbia University, a Bachelor of Science in Electrical Engineering and a Bachelor of Arts in Philosophy from the University of Maryland, College Park. He is also founder and president of The Xavier Zubiri Foundation of North America. He serves as editor of two journals, *The Telecommunications Review* and *The Xavier Zubiri Review*. He is the author of over 100 articles, papers, and reviews. He has translated two books, and has given courses and papers in the United States, Canada, Mexico, South America, and Europe. Evolution, especially its systems and mathematical aspects, has been a subject of lifelong interest to him. Together with Daniel Kuebler, he published *The Evolution Controversy* in 2008, the only major book analyzing the issues surrounding evolution in an objective manner. Dr. Fowler is a member of several honorary societies including Phi Beta Kappa, Eta Kappa Nu, and Tau Beta Pi. He is a Senior Member of Institute of Electrical and Electronics Engineers (IEEE), and a member of The American Mathematical Society (AMS), the American Physical Society (APS), and American Association for the Advancement of Science (AAAS).

Julio A. Gonzalo has a PhD in Physics from the Universidad Complutense, Madrid. He did research and teaching at the universities of Salamanca (Spain), Mayagüez (Puerto Rico), Rio Piedras (Puerto Rico), Barcelona (Spain) and UAM Madrid (Spain). He worked as Research Collaborator for the US Atomic Energy Commission at Brookhaven National Laboratory, Upton, L.I., New York (1963–64) and as Division Head and Sr. Scientist at the PRNC-USAEC (1965–76). Now he is Professor Emeritus at the Universidad San Pablo-CEU, Madrid. He has been author or co-author of over 200 publications and a dozen scientific books, the latest is *Cosmic Paradoxes* (World Scientific: Singapore, 2012).

Lucía Guerra-Menéndez holds a Licentiate in Pharmacy (2005) and a PhD in Medicine (2011) by the University San Pablo-CEU, Madrid. She has also a Diploma of Advanced Studies in Bioethics and Bioethical Legislation (UNESCO), a Master Business Administration (MBA) by IEB, and has presented invited oral communications on Science and Religion at Philadelphia (2005–7), Madrid (2007–8), Mexico (2008) and Rome (2008–9). At present she is Assistant Professor at the Universidad San Pablo-CEU, Madrid.

Nicolás Jouve de la Barreda obtained the PhD degree in Biology at the Universidad Complutense, Madrid (1973). He is chair professor of Genetics. Currently he teaches courses of "Genetics" (Schools of Medicine and Biology), "Evolutionary Genetics" and "Evolution of Genes and Genomes" (School of Biology) at the University of Alcalá. He was awarded academic recognitions by the Social Council of the University of Alcala for his "Scientific Research" (1991) and "Teaching" (1996) activities. He was invited to teach different short courses by the Universities of La Serena (Chile, 1996), Mendoza (Argentina, 1998) and León (Nicaragua, 2001), and was a guest researcher in the University of Missouri, Columbia, MO, USA, (1988). He was President and re-founder of the Spanish Society of Genetics (1900 to 1994). He is author of more than 200 papers in prestigious journals of his specialty (Cytogenetics, Genetics and Molecular Biology of cultivated plants). He is author of several books: *Genetics* (Ed. Omega, Barcelona, 1989, 3 editions), *Biology, Life and Society* (Ed. Antonio Machado, Madrid, 2004), *Exploring the Genes. From the Big-Bang to the New Biology* (Ed. Encuentro, Madrid: 2008, "two" editions), *The Spring of the Life. Genes and Bioethics* (Ed. Encuentro, Madrid: 2012).

Daniel Kuebler received his PhD in Molecular and Cell Biology from the University of California, Berkeley in December of 1999 and earned a Masters of Science in Cell and Molecular Biology from the Catholic University of America in 1995. In addition, he has a Bachelors's degree in English also from the Catholic University of America.

For the past 11 years he has been a Professor of Biology at Franciscan University of Steubenville where he has been responsible for teaching courses in evolution, cell biology and human physiology. His academic research involves an investigation of the physiological defects of seizure disorders and his work has been published in journals such as *Brain Research, Genetics and the Journal of Neurophysiology.*

He is the co-author of a book entitled *The Evolution Controversy: A Survey of Competing Theories* (Baker Academic), which is a scientific critique of the various theories of evolutionary thought. He has also presented and published academic papers regarding evolution and faith at a number of national and international meetings. In addition to his academic work, he has authored a variety of popular articles on science, evolution, politics and religion that have been published in places such as the *National Catholic Register* and the on-line journal *Public Discourse*.

Fernando Sols is Professor of Physics at Universidad Complutense de Madrid. He graduated from Universidad de Barcelona (1981) and took his PhD at Universidad Autónoma de Madrid (1985). He has been Fulbright Fellow at the University of Illinois at Urbana-Champaign, Associate Professor at UAM, Director of the Nicolás Cabrera Institute at UAM, and member of the Editorial Board of the *New Journal of Physics* (IOP-DPG). He is Fellow of the Institute of Physics (UK). At present he is Director of the Materials Physics Department at UCM. He conducts research on theoretical physics problems related to the dynamics and transport of electrons in solids and cold atoms in optical lattices, and has a general interest in quantum dissipation and macroscopic quantum phenomena.

Foreword

The monumental achievements of modern science, from elementary particles to atomic physics, from geophysics to cosmology, from molecular biology to genetics, attests to intelligibility of the world around us.

This implies, logically and metaphysically, an intelligent Creator. No random, blind, creative force of any kind can explain itself or the complexity of modern science.

In this book a team of distinguished authors and scientists approach from diverse and often complementary viewpoints the evidence for intelligible design in the world around us: Julio A. Gonzalo (Professor of Materials Physics, UAM Madrid), Manuel M. Carreira, S.J. (Professor of Astrophysics, Vatican Observatory), John Beaumont (Lawyer, Leeds, England), Manuel Alfonseca (Professor at the Politechnique School, U. Autónoma, Madrid), Thomas B. Fowler (ScD, George Washington University), Daniel Kuebler (Franciscan University of Steubenville), Nicolás Jouve (Professor of Genetics, U. Alcalá de Henares), and Lucía Guerra Menéndez (Assistant Professor, Faculty of Medicine, U. San Pablo CEU, Madrid).

In the first part the authors deal with the origins of modern science in historical perspective, the limits of science, and the intelligibility of the physical universe.

In the second part, they concentrate on the origin and development of life in general and human life in particular. This part starts with a brief history of evolutionary thought, includes the contributions of genetics and the human genome to the understanding of life, and deals with the evolution controversy.

Finally, they deal with such questions beyond nature and history and briefly with the questions of free will.

The most recent Gallup poll (2012) of Americans on human origins shows the following responses to the question: Which of the following statements comes closest to your view on the origin and development of human beings?

1. Human beings have developed over millions of years from less advanced forms of life, but God guided the process (Response: 32%).
2. Human beings have developed over millions of years from less advanced forms of life, but God had no part in the process (Response: 15%).
3. God created human beings pretty much in their present form at one time within the last 10,000 years or so (Response: 46%).

This seems to imply that, at least for America and for the time being, a sizeable majority of those responding see a Creator beyond our intelligible universe.

In this book, sections 1.1 and 1.7 (JAG), 1.2, 2.6 (MMC) and 2.1 and 2.5 (TBF) are adapted from previously published work by the respective authors.

We are indebted to M. de la Pascua for valuable suggestions regarding the text and the book's illustrations.

The Editors

Part I. Modern Science in Historical Perspective

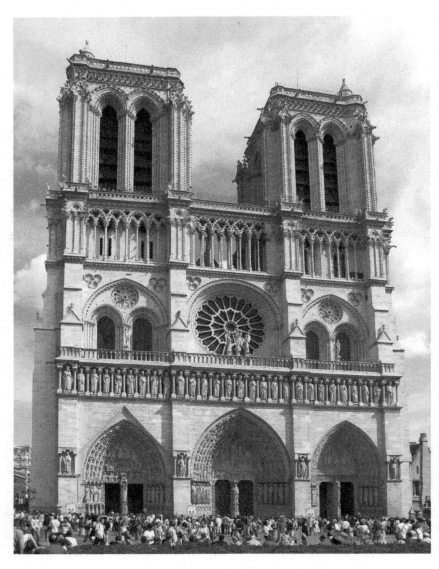

Notre Dame de Paris: Western Facade

1. On the Origins of Modern Science

Julio A. Gonzalo

Like the origins of many branches of learning, the origins of historiography can be traced to early Greek authors: Herodotus (487?–425?), Xenophon (430?–355?) and Polybius (203?–120?), all of them consummate analysts and narrators.

The question "What is the meaning of history?" may make sense for modern Western men, but not for Polybius, who better formulated the question. In telling the story of the Roman conquest of Greece, he put the future on the side of the Romans (who in his eyes were nothing but a nation of barbarians), though only for a while. According to Polybius, the Romans would ultimately go down and be defeated, because no dominant society at any time is above the most basic law of history. For Polybius, and for the ancient Greeks, this law was the *law of the wheel*: the inevitable sequence of events in which birth, progress, decay and death follow each other endlessly in human affairs.

In Sanskrit the word *swastika* is a compound of the words *su* and *asti*, which mean, respectively, *well* and *being*. It is curious that symbol (the swastika) was most popular with the Greeks (then decaying) and with the Romans (then on the rise) when Polybius wrote. When the Nazis adopted the swastika as their emblem, they were consistent in saying that the New Europe at which they arrived would not last for ever.

Many centuries later, the Chinese, who had to their credit the inventions of block-printing, gunpowder and magnetic needles (Francis Bacon wrongly saw them as the beginning of science) could have sailed East carrying small

3

cannons to colonize the new world before the Spaniards arrived to Western Mexico and California.

This could have happened, as S. L. Jaki points out, if the Chinese had been truly scientific, in the sense of being able to develop their inventions in a systematic and consistent manner. History might have been very different in that case.

But history became global only in the sixteen and seventeen centuries when it became Western history, first with the ships of Spain and Portugal, then with those of England, Holland and France. History, for the last three hundred years has been on the side of the West, and, until recently, serious challenges to Western global domination, like those of Soviet Russia and Communist China, have been based mostly upon the availability of western technology. Only recently, with China catching up with Western science and technology is this beginning to seriously challenge Western superiority. At the same time the West, both America and Europe, have begun to suffer internal problems of their own that weaken their prospects of superiority in the future.

The development of a fully advanced science, however, has taken place only once in world history. When Aristotle's work became known in full in the West, in the late thirteen century, Western Europe was deeply Christian. She was known as *Christendom*. Then, at the newly established universities, Dominican and Franciscan teachers, among others, thought everything knowable ("omnis res scibile") from a truly Christian perspective. The Catholic Church was the driving force. The past history of the world was finite, in this perspective, and the world itself, rather than eternal and infinite, was finite, open and contingent. This was something inconceivable in the Greek perspective. The idea that everything in motion had to have an absolute starting point was then and there the natural starting point of science, a science in which one discovery generates another discovery, and science becomes a self-sustaining intellectual venture.

As we will point out below, Jean Buridan, professor of natural philosophy at the Sorbonne of Paris around 1330, can be considered the first modern physicist. He introduced the concepts of *inertial motion* and *momentum*, which paved the way, through Copernicus, Galileo and Kepler, to the "*Principia Mathematica*" of Newton which, by setting forth a definitive text of Mechanics, opened the way to future developments in Optics, Electricity, Thermodynamics, and Modern Physics.

In his commentaries to Aristotle Buridan voiced a profound disagreement with the Philosopher. He rejected the notion of the eternity of the universe and of all motion in it, including the motion of the stars, as incompatible with the fundamental truths of Christian Revelation. He also offered a penetrating speculation about the motion of celestial bodies. Buridan formulated then and there Newton's first Law of Motion, which is tantamount to making the first step, conceptually, but also historically, in modern science. His statement was copied in countless manuscripts, which Buridan's students carried all over Europe, from Salamanca to Krakow. This statement, or its equivalent, according to Stanley L. Jaki, was well known to Galileo and Descartes, who are still credited with formulation of the first of Newton's Law by many contemporary historians of science. The rediscovery of Buridan's first step in modern science was made at the beginning of the 20th century by a relatively young French professor at Bordeaux, Pierre Duhem, then 45 years old and a widower, who already had an international reputation as expert in thermodynamics and continuous mechanics. Until then Duhem had taken for granted that there had been no "science" in the Middle Ages, and that the theological mentality at that time was hostile to creative scientific thought. This derived in part, from the hostility of the leaders of the French Enlightenment to anything Christian. But when he traced the concept of virtual velocity further and further back in history, beyond Galileo to his teachers, Benedetti and Stevin, and then to Cardan, while investigating the latter's cryptic reference to a certain Jordanus, Duhem discovered the work of Jean Buridan and his disciple Nicole Oresme.

The simplistic popular notion that science is hostile to religion, fueled by not a few well written popular scientific books (those of Sagan, Asimov, etc) is simply not true. Exactly the contrary: modern science had its origin in medieval Christian Europe, as meticulously documented by Pierre Duhem (1861–1916) in his monumental work *"Le systeme du monde"* (A. Hermans et Fils: Paris, 1913...), and further expounded and developed in numerous books and publications by Stanley L. Jaki (1924–2009), one of the foremost historians of science of the twentieth century.

As Jaki recounts in detail in *The Origin of Science and the Science of Its Origin*, some enlightened European freethinkers who were contemporaries of Voltaire, after reading extensive reports on China by Jesuit missionaries, were impressed by Chinese achievements in ethics and moral philosophy, architecture, engineering, arts and crafts, but very little in science. The work of Fr. Louis Lecompte SJ, "Nouveaux Memoires sur l'état present de la Chine" (1696) was composed of fourteen long letters to various civil and ecclesiastic French dignitaries. It was soon translated into English, German and Dutch, and covered many topics on Chinese geography, politics, history, literature, arts and crafts, as well as science. The science that Fr. Lecompte had in mind to make a comparison with Chinese science was Euclid's and Ptolemy's science. However, at the time, European science had just achieved full maturity in the *Principia Mathematica* of Newton. This maturity had its roots in medieval Christendom, in the seminal work of Jean Buridan and Nicole Oresme, whose pioneering ideas provided fertile ground for the decisive developments of Copernicus, Galileo and Kepler and, finally, Newton.

In view of the considerable talents abundantly demonstrated by the Chinese through their millenary history, their achievements in science were certainly meager, compared with those in Europe at the time.

Then the question: *Why not in China?*

But the proper question to be asked is not the one made by those European free thinkers: That question should have been: Why science, a self-sustaining science, worthy of that name, had developed only in one cultural matrix, the cultural matrix of *Medieval Christendom?*

Other great civilizations in world history could be justly proud of their stupendous achievements in architecture, public works, arts and crafts, drama, literature, even in philosophy and logic, but not in science proper, or at least in any degree comparable to the level achieved at the beginning of the eighteenth century in Europe — an achievement with roots in Medieval Christendom as shown by Duhem and Jaki.

In the period going from the early twelfth century to the time of Buridan and Oresme, during which the concept of "impetus" and the concomitant idea of "inertial motion" were introduced for the first time, one can see developments which lead directly to the formulation of the fundamental laws of motion (Newton's laws), developments connecting in one stroke motion here on Earth and motion in the sidereal realm: the motion of the "planets" or "wanderers".

Slowly at first, then at a fast pace, Copernicus, Galileo and Kepler, prepared the way for Newton.

Abelard of Bath (c. 1125) embarks on long and arduous journeys in the quest of learning, going as far as the Middle East, and brings to medieval Europe the elements of trigonometry, the art of making astrolabes and Euclid's geometry. His contacts with Muslim learned men made him aware

7

of the ongoing struggle in the Arab culture to reconcile faith and reason. Abelard is on record remarking to his nephew that many of his contemporaries, including Muslim and Jewish men of learning, identified God with Nature. There was, at this time and at any other time, a strong tendency to identify nature with the ultimate entity. This tendency can be overcome only if men are willing to recognize their dependence on a truly transcendent Creator. Abelard favors, whenever possible, natural explanations over miraculous ones, showing his healthy proclivity to a true scientific attitude when facing the physical world: "I do not detract from God... Whatever there is, it is from Him and through Him. But the realm of being is not a confused one... Only when reason totally fails should the explanation of the matter be referred (directly) to God".

In other words, Abelard sees in Nature Nature's God without any need to deny the world of the supernatural, which, if real, is also God's. Medieval men were tempted too by the mirages of fatalism and astrological pantheism.

Thierry of Chartres (d.c. 1155) rises well above the Greek animism and pantheism, latent even in the best literary exponents (such as Plato's *Timeus*) saying: Moses' intention was to show that the creation of all things and the formation of men was made by the only one God to whom alone worship is due. The usefulness of [Moses'] works is the acquisition of knowledge about God through His handiwork". For Thierry "there are four kinds of reasons that lead man to the recognition of his Creator: the proofs are taken from arithmetic, music [harmony], geometry and astronomy". If the Creator has actually arranged everything "according to number, measure and weight", as recorded in the Book of Wisdom, man's intellectual understanding of the world has to have a mathematical, scientific character.

Robert Grosseteste (c. 1168–1253), possibly the first chancellor of Oxford University, was even more explicit about the mathematical understanding of nature: "The usefulness of considering lines, angles and figures is the greatest, because it is impossible to understand natural philosophy without them. They are efficacious throughout the universe and its parts and the properties related (to them), such as rectilinear and linear motion". Grosseteste's investigation of the rainbow is a good example of his scientific methodology, which included seminal programs of induction, falsification and verification. He rightly attributes the rainbow to light's refraction rather than to its reflection, as done before incorrectly by Aristotle and Seneca. Grosseteste's methodology depends on the idea of the Creator as a wholly rational and personal Planner, Builder and Maintainer of the Universe.

William of Auvergne (d.c. 1249) in *De Universo* makes fragmentary references to magical and astrological aphorisms but he does not succumb to the irrational in his quest for understanding, and disputes continuously with "Manichaeism, fatalism, pantheism, star worship" and similar betrayals of man's rationality, which he associates with the Greeks of old, with the Saracens and with the hermetic philosophers. In his lengthy discussion of the Great Year, identified with the period of 26,000 years needed to complete a precession of the equinoxes, William of Auvergne was aware of the belief in "eternal returns", which he rightly sees as the embodiment of a pagan, non-Christian worldview. Defending right reason against the Mutakallimun (devout and learned Muslim philosophers), he points out the fatal error of not distinguishing clearly between primary and secondary causality.

Thomas Aquinas (1225–74) embarks in a gigantic effort to bring together *faith* and *reason* "in a stable synthesis". His theory of knowledge is moderate realism and the doctrine of the analogy of being becomes the key of his metaphysics. His resolute commitment to give reason its due meant a generous acceptance of the Aristotelian system, then for almost two millennia the epitome of a rational explanation of the world. It was motivated (according to Stanley Jaki) by his attention to contemporary Muslim theologians and philosophers. His "*Summa contra gentiles*" is intended to counter the occasionalism and the fatalism contending with each other within the Muslim contemporary culture. His "Summa *Theologica*", a work of synthesis, aims to show that the reason for the existence of the cosmos is its subordination to man's supernatural destiny. Aquinas, contrary to his master Albertus Magnus, is not interested in experimental investigation, but both disciple and master agree in the all important point of rejecting the inevitability of "eternal recurrences" in the world. In his "*De facto*", Albertus is most explicit about the history of the question of "eternal returns" in Plato, Aristotle, the Stoics, Ptolemy, the Arabs (Albumasar especially), all of them inclined in favor of "eternal recurrences". The early Church Fathers were opposed to this idea as well.

In his famous Five Ways, Aquinas, beginning with specific characteristics of actual existing beings, arrives at the necessary existence of a Prime Mover, an Uncaused Cause, a Necessary Being, a Most Perfect Being and an Ordering Intellect, who everyone understands to be God.

It may be noted that, towards the end of the eighteenth century, enlightened philosophers began to take as an indisputable fact that there is no way to go from the "cosmos" ("a bastard product of the metaphysical cravings of man" according to Kant) to the Creator. Thanks to Einstein's General Theory of Relativity, we now have a contradiction-free concept of the "cosmos", a finite "cosmos", which is a perfectly valid ground to go from its existence to the Creator's existence.

Let us quote Jaki on Chesterton to that effect:

"No wonder that (Chesterton)... spoke devastating words of that philosopher, Kant, who more than any other succeeded in leading mankind into the belief that the universe was the bastard product of metaphysical cravings: Long essays on Kant and the German idealists contain far less than these few words of Chesterton: 'The note of our age is a note of interrogation. And the final point is so plain; no skeptical philosopher can ask any question that may not equally be asked by tired child on a hot afternoon. Am I a boy?-Why am I a boy?-Why aren't I a chair?-What is a chair? A child will sometimes ask questions of this sort for two hours. And the philosophers of Protestant Europe have asked them for two hundred years.' ..."

(See Stanley L. Jaki, "Chesterton, a Seer of Science",
University of Illinois Press Urbana and Chicago, 1986)

Roger Bacon (1214–94). It seems that Bacon's impetuous efforts to secure the service of science for the Christian faith led him to compose his *Opus majus*, which resulted in his temporary imprisonment. Apparently, his novel views, stressing too much the inexorable determinism of events in nature, made them at the time sound incompatible with man's freedom and moral responsibility. But, according to Jaki, Bacon did not capitulate, as most Arab commentators of Aristotle did, to the idea of cyclic determinism, and he made a clear distinction between man's supernatural destiny and his earthly everyday existence, between final (primary) causes and efficient (secondary) causes. For him man's knowledge about nature was always partial, not "a priori" in character. One of the most learned and celebrated teachers at Oxford in his time, he was a scientific pioneer in controlled experiment and accurate observation of natural phenomena. He said that mathematics was the gateway to science and experience, and verification the only basis of certainly.

Etienne Tempier (Bishop of Paris), in 1277, condemned 219 propositions mostly against Siger of Brabant and his followers, including the eternity of the world (p. 83–91) and the perennial recurrence of everything every 26,000 years (p. 92). He was affirming in substance the rigorous contingency of the world with respect to a transcendental Creator, source of all rationality on heaven as well as on Earth. Pierre Duhem, the great French physicist and historian of medieval physics, takes the decrees of Bishop Tempier as the starting point of a new era in scientific thinking. In the decrees, the possibility of several worlds was recognized in principle (p. 27); rejected, on the other hand, was the existence of animated, incorruptible and eternal superlunary bodies (p. 31–32); the possibility of rectilinear motion for celestial bodies as part of a celestial machinery was admitted (p. 75); rejected was the deterministic influence of the celestial stars on the lives of individual men from the instant of their birth (p. 105); likewise rejected was the provenance of a "first matter" from celestial matter (p. 107), etc. All of this was aimed at defending the freedom and the exclusive rights of the Creator when he created Heaven and Earth.

The bishop's statements were not binding in the universal Church but were intended to uphold orthodoxy in the University of Paris where Siger of Brabant had been teaching for more than a decade.

As pointed out by Alfred North Whitehead in 1925 in his *Lowell Lectures*:

"I do not think however that I have even yet brought up the greatest contribution to the formation of the scientific movement. I mean the inexpugnable belief that every detailed occurrence can be correlated with its antecedents in a perfectly definite manner, exemplifying *general principles* (emphasis added). Without this belief the incredible labours of scientists would be without hope. It is this instructive conviction, vividly posed before the imagination, which is the motive power of research: that there is a secret, a secret which can be unveiled. How has this conviction been so vividly implanted in the European mind!"

Jean Buridan (1300–58). He, according to Stanley Jaki, through his commentary on *De Caelo* by Aristotles, had an unmistakable influence on Galileo. His "Questiones super quattor libris de caelo et mundo", through the slightly modified version by his disciple Albert of Saxony, were commonly available at medieval European universities from Salamanca to Krakow. The Aristotelian clear-cut distinction between superlunary and sublunary matter was dealt a decisive blow in Buridan's work. He reminded his readers that the heavens could decay and that the Creator was perfectly free to amihilate the world if He so wished: "In natural philosophy one should consider processes and causal relationships as if they always come about in a natural fashion; God is no less the cause, therefore, if this world and its order have an end, than if this world is eternal". Buridan did not accept yet the rotation of the Earth (his disciple Oresme later did) but, against Aristotle, he held that continuous, though relatively small, changes between the Earth and the fixed stars were possible.

Buridan proposed, against the mistaken Aristotelian notions on motion, his new concept of "*impetus*", a quality implanted in the moving body by the mover, and his notion of "gravity", a property innate to all massive bodies. And after reviewing the usefulness of his new theories to describe various motions here on Earth, he dared to outline their usefulness for celestial mechanics. He wrote, in the same breath, about a jump and about ordinary planetary motion, with the Creator as the ultimate agent imparting a given quantity of motion to the various parts of the universe: "He (God) created the world, moved all the celestial orbs as He pleased, and in moving them He impressed in them "impetuses" which moved them without his having to move them any more except by the ... general influence whereby He concurs as co-agent in all things which take place... And these impetuses which He impressed in celestial bodies were not decreased or corrupted afterwards because there was no inclination [to it]... Nor there was resistance which would be corruptive or resistive of that impetus... But I do not say this assertively, but [rather] tentatively..." [For a more complete discussion see Stanley L. Jaki, "*Science and Creation: From eternal cycles to an oscillating universe*" (Lanham: New York, 1990) and references therein].

Nicole Oresme (1323–82). The most outstanding disciple of Buridan, wrote with great originality on a variety of topics, including monetary theory, astronomy and algebra. His commentary on Aristotle's *De Caelo* is considered today a classic of early scientific literature. For Oresme the perfection of the laws of nature is but a modest reflection of the infinitely perfect attributes of the Creator. Oresme allowed corruptibility in the celestial motions of celestial bodies only in the restricted sense that these motions were frictionless. Against the possibility of eternal recurrences, he noted that the periods of the planets are most likely incommensurable. He pointed out that, in any case, such coincidences should occur after periods much larger than 26,000 years, or the period of the precession of the equinoxes (the length assigned in ancient Greece to the "Great Year"). Oresme parted with Aristotelian necesitarianism, which implied that several worlds would require several Gods. He responded: "One God governs all such worlds".

Many years later, "*impetus*" would be correctly redefined as "*momentum*", i.e. the product of inertial mass times the velocity imparted to the moving body. The full process from Buridam and Oresme to Newton took three hundred years. **Copernicus** (1473–1543), with the "impetus" theory behind him, gave the crucial step of postulating the heliocentric description of planetary motion. **Galileo** (1564–1642) and **Kepler** (1571–1630) did the next important steps. A Christian worldview was essential in the whole process.

How did Christian belief provide a "cultural matrix" for the growth of science?

Stanley Jaki in his booklet on "Christ and Science" gives four reasons:

1. "... the Christian belief in the Creator allowed a breakthrough in thinking about nature. Only a transcendental Creator could be thought of as being powerful enough to create a *nature* with *autonomous laws* without his power over nature being thereby diminished. Once the basic among those laws were formulated, science could develop on its own terms".

2. "The Christian idea of *creation* made still another crucially important contribution the future of science. It consisted in putting *all material beings on the same level* as being mere creatures... The assumption would have been a sacrilege in the eyes of any one in the Greek pantheistic tradition or in any similar tradition in any of the ancient cultures".

3. Finally, *man* figured in the Christian dogma of creation as being specially created in the image of *God*. This image consists in man's rationality as somehow *sharing in God's own rationality* and in man's condition as an ethical being with eternal responsibility for his actions. "Man's reflection on his own rationality had therefore to give him confidence that his created mind *could fathom* the rationality of the *created realm*".

4. "At the same time, that very createdness could caution man to guard against the ever-present temptation to *dictate to nature what it ought to be*. The eventual rise of the experimental method owes much to that Christian matrix".

In contrast to those Christian natural philosophers, who brought about the birth of science in Medieval Europe, some of today's leading theoretical physicists try to dictate to nature what it ought to be.

Nowadays many educated people believe, only because they have been culturally conditioned to think so, that Christianity — and very specially, the Catholic Church — are and have always been hostile to science. Nothing is further from the historical truth.

Modern science is characterized by an impressive capability to describe in quantitative terms an enormous variety of natural facts. From elementary particles to galaxies, to the universe as a whole through differential equations and quantitative mathematical solutions, it is capable of matching the most precise experimental data. In dynamics there is a subtle continuous development[3] from Buridan to Copernicus, from Copernicus to Newton, from Newton to Einstein. In electrodynamics there is also a subtle continuous and logical development from Coulomb to Volta, to Ampere, to Faraday to Maxwell. And in thermodynamics there is continuous development from Carnot, Joule and Mayer to Clausius, Kelvin and Planck. As physicists, for all of them, the working of the physical laws is coherent and consistent, and the working of man's intellect, sometimes very laboriously, is capable of discovering order and harmony in nature.

In our contemporary Western civilization, a civilization with roots in medieval Christendom, generation after generation of curious observers and a handful of geniuses have made a continuous progress and a series of momentous discoveries, always following a middle road between shortsighted positivism and an aprioristic idealism.

The intellectual adventure of building a scientific corpus in the Christian West was neither a random development nor a deterministic one, obviously.

If the world had not been made rationally, systematic scientific knowledge would be impossible. If something rationally valid here and now could be false at another place and another time, a substantial effort to investigate and study it would not be justified; physics as well as metaphysics presuppose the existence of an objective reality independent of the observer.

It does not require a scientist to see the truth that the world is rationally made. The astonishing edifice of contemporary science is proof that the laws of nature can be quantitatively and systematically described. They are, therefore pointers to the wisdom of the Creator of that nature and that laws which govern it. At the same time, science is a monumental proof that man has been endowed by the Creator with intelligence and freedom to investigate nature and nature's laws. The Christian West is the heir of Jewish biblical wisdom and of early Greek and Roman intellectual achievements. After the decline and fall of the Roman Empire, physical science suffered many delays and detours to be born finally in a Medieval Christian matrix. In this matrix, at long last, cosmic "eternal returns" and the pagan doctrine of the Great Year could be left behind.

At the turn of the nineteenth century and in the first years of the twentieth, Pierre Duhem, who had made already great contributions to theoretical physics in mechanics and thermodynamics (and had written a penetrating philosophical interpretation of physical laws) discovered, to his own surprise, the medieval origins of Newtonian physics at the University of Paris in the first half of the fourteenth century.

Duhem's extraordinary contribution has been somewhat reluctantly recognized in secularist academic circles but has been largely ignored in nominally Christian university campuses. In his book[4] *Scientist and Catholic: Pierre Duhem,* Father Jaki gives a vivid portrayal of the dramatic life and work of Pierre Duhem. In the second half of the book he offers a representative sample of selections of Duhem's writings which illustrate well the unity accomplished in him of his science and his Catholic faith. R.P. Feynman,[5] not a believer himself, said something which undoubtedly applies to Pierre Duhem: "many scientists do believe in both science and God — the God of revelation — in a perfectly consistent way". Historical perspective will keep competent physicists as far away from radical dogmatism as from radical skepticism.

In his "Le systeme du monde... Tome II. La Cosmologie hellenique (1914), pp. 390, 407–408 (1914–1)" Duhem summarizes the role of the Medieval Catholic Church in destroying the pagan doctrine of the "Great Year" which, of course, implies an eternal universe, difficult to reconcile with the Big Bang.

"In the system which Maimonides sets forth we see, so to speak, the culmination of all the ideas whose development has been traced in this chapter.

We find there, first of all, the affirmation of the principle that Aristotle had already formulated with such clarity: The various parts of the universe are interconnected by a rigorous determinism and this determinism subjects the entire world of generation and corruption to the rule of celestial circulations.

We find there the corollary of that principle, namely, the definition of an astrological science which ties all changes accomplished here below to the motion of a specific planet.

We see there the preponderant role which that astrology attributes to the Moon as a rule of water and humid matter. The Moon forces them to grow and decrease with her. The theory of tides clearly proves the reality of this lunar action and, through it, of all influence emanating from the celestial bodies.

Finally, we hear stated that the very slow changes on earth are tied to the almost imperceptibly slow motion of the fixed stars whose revolution measures the Great Year.

To that system all disciples of Greek philosophy — Peripathetics, Stoics, and Neoplatonist — have contributed.

To that system Abu Masar offered the homage of the Arabs. The most illustrious rabbis, from Philo of Alexandria to Maimonides accepted that system.

Christianity was needed to condemn that system as a monstrous superstition and to throw it overboard...

Hardly anxious to explore in detail the works of Greeks astronomers, the bishop of Hippo and with him, undoubtedly, the great majority of the Church Fathers, did not know how to separate, in a precise manner, the hypotheses of the astronomers from the astronomers' superstitions. The former were confusedly included in the disapprovals accorded to the latter...

Let us not therefore search in the writings of the Church Fathers for the traces of a meticulously and sophisticatedly treated science. We assuredly cannot find them there at all.

Let us not, however, neglect the little they said about physics and astronomy.

The first of their teachings on this topic are the first seeds from which the cosmology of the Christian Middle Ages would slowly and gradually develop.

Also, and above all, the Church Fathers hit, and did so in the name of the Christian Creed, the pagan philosophers on points which, today, we consider more metaphysical than physical but where actually lie the cornerstones of the physics of Antiquity: Such are the theory of eternal prime matter, the belief in the stars domination over sublunary things and in the periodic life of a cosmos subject to the rhythm of the Great Year. By destroying through these attacks the cosmologies of peripatetism, of Stoicism and of Neo-Platonism, the Fathers of the Church clearly prepared the way for modern science".

As pointed out above, one can trace[6] the continuous development of physics, or the science of massive bodies in motion, from Buridam to Copernicus, and from Copernicus to Newton.

Copernicus leaned about Buridan's ideas in his critical commentaries to Aristotle's cosmological work, *On the Heavens*, which he studied at the University of Krakow, where the university library still has a dozen copies of Buridan's manuscript. Such copies can be found in many other significant European medieval universities. Buridan's ideas were further developed by his disciple, Nicole Oresme, who succeeded him at the Sorbonne.

The passage previously quoted of Buridan[7] makes clear its Christian theological matrix: belief in creation out of nothing and in time. This belief was held explicitly from early patristic times, and it was defined in 1215, at the Fourth Lateran Council. Both Aristotle and Ptolemy, as all other scholars of classical pagan times, were eternalists. The world had no beginning and no end, and, if it had large-scale changes they were only for the duration of a Great Year, or the duration of the full precession of the equinoxes, about 26,000 years.

Today, three hundred years after Newton, inertial motion looks very natural. Why did that idea come so late? To answer this question it is necessary to take a look to Aristotle's ideas on motion. According to him the heavenly sphere moves because it is animated by its desire for the Prime Mover. But it would be a mistake to take Aristotle's Prime Mover for a Creator transcendent to the universe. As pointed out by Fr. Jaki, Aristotle spoke often as a pantheist for whom the universe was not only the Supreme Being, but the supreme living being.

The "animization" of the world is the essence of Plato's and Aristotle's systems, the two best followers of Socrates. It was very difficult if not impossible to formulate the laws of motion in nature, beginning with the law of inertia, in an "animated" world. So, according to the penetrating analysis of Fr. Jaki, the coming of modern physics (anticipated as shown above by Buridan, and then by Copernicus, Galileo and Kepler) was delayed until a Christian theological matrix could produce an intellectual climate according to which nature was created and was not the ultimate and supreme living being.

As a result, physical science witnessed only one viable birth: in Western medieval Christianity.

The medievals,[8] and this may surprise many, made tremendous advances in producing technical devices. Until the advent of the steam engine, the Western world lived on technological innovations made during the medieval centuries. Among these innovations one should mention the *cam*, which allows the transformation of circular motion into linear motion and vice versa (therefore, the power provided by watermills could be used to drive mechanical saws, etc.) and the transformation of accelerated motion to motion at *constant velocity* (which made possible the construction of pendular clocks by the end of the thirteenth century, a vast improvement for time measurements).

For the medievals, the Christian belief in the Creator allowed a breakthrough in thinking about nature. Only a truly transcendental Creator could be powerful enough to create a nature with autonomous laws without his power being thereby limited.

Unlike the pagan Greek cosmos, there could be no divine bodies in the Christian cosmos. All bodies, heavenly and terrestrial, were now on the same footing. This eventually made it possible to think that the slow fall of the Moon in its orbit and the fall of an apple on earth could be governed by the same law of gravitation.

References

1. S.L. Jaki, *Science and Creation* (University Press of America: Lanham, MD, 1990).
2. S.L. Jaki, *Christ and Science* (Real View Books: Royal Oak, Michigan, 2000).

3. Julio A. Gonzalo, *The Intelligible Universe* (Singapore: World Scientific, 2008).
4. Stanley L. Jaki, *Scientist and Catholic: Pierre Duhem* (Front Royal, VA: Christendom Press, 1991).
5. Richard P. Feynman, The Relation of Science and Religion, *Engineering and Science* 19(9) (June, 1956), p. 20.
6. Stanley L. Jaki, *Christ and Science* (Royal Oak, Michigan: Real View Books, 2000) p. 13.
7. Ibidem. p. 15.
8. Ibidem. p. 22.

This section has been adapted from previously published book by the author in Julio A. Gonzalo and Manuel M. Carreira, *Everything coming out of nothing* (Bentham Science, 2012).

Nicolaus Copernicus (19 February 1473–24 May 1543)

2. The Post-Renaissance Revolution: The New Science

Manuel M. Carreira, SJ

Universidad Pontificia de Comillas

During a time period centered between 1550 and 1700, a new way of thinking about material nature was developed in post-Renaissance Europe. This new approach, that embraced an ever increasing body of theoretical and technical knowledge, forms the framework of Modern Science. It is the outcome of a new "model", a viewpoint that made possible — through apparently insignificant advances — the attainment of the old dream of understanding the Universe by finding logical reasons for its structures and interactions both at the astronomical and at the microscopic levels, something that the more ambitious efforts to attain a single abstract explanation had failed to provide.

Perhaps the deepest reason — not always consciously pursued — was that the entire material reality ceased to be considered as a single whole, made and structured once for all, accepting instead that there was and is evolution both in "inert" and in living matter. This way of thinking implies the need for interactions, activities that are ruled by "laws" that are simple statements of how things occur due to the nature of matter, not to any kind of external imposition. This reasoning process is applied in Astronomy, Geology, Physics, Chemistry, Biology with consequences of enormous theoretical and practical impact that establish the basis for the industrial revolution and that change the approach to Medicine, to the Economy, to the transmission of culture. Thus we find the new intellectual atmosphere as something we now accept as "obvious" but that we really must admire, mostly for its rapid development within human history.

In order to better understand this entire process it will be useful to remember how the way of thinking evolved within Western civilization, without forgetting that in other areas — the Orient especially — important steps were taken as well, mostly of a technical order. We cannot give a complete historical account, but we shall underline moments and personal contributions that appear as critical through more than twenty centuries. Thus we can specify some models or viewpoints for which we have historical references.

Mythological and religious models

In ancient civilizations for which we find sufficient archaeological evidence to infer their way of thinking regarding the material world (Babylon, Egypt, India, the peoples of central America among others) we uncover an early period when the human experience of the world is interpreted along polytheistic religious ideas. We can't always clearly distinguish pantheism from polytheism and animism: in all of those there is a basic thought that accepts a kind of divine character for everything outside the ordinary experience of human life (even if an existence after death somehow changes also our own nature into something supernatural).

The material world — apparently quite inert — is really considered as endowed with some mysterious free and vital activity tied to it: there are spirits or gods of the air. The fire, the rivers, the mountains, the trees, the sea, the storms. Even animals, especially those that are either dangerous or most useful, are in some way divine or linked to a divinity, the snake, the bull, the jaguar, the eagle... The divine dignity is especially assigned to those things that are inaccessible or impressive and mysterious: the Sun and the Moon, the planets, volcanoes, the Earth itself...

Activities in nature are not considered as due to properties of a matter endowed with forces and fixed ways of acting: free superhuman beings determine everything that happens from the daily sunrise to the unexpected turns of human life. Instead of Astronomy we find Astrology: an effort to interpret the plans of the gods for each individual, and to use magic to control future events that lack all sense of order and security. In many instances there is a background of rivalries among those entities that control the world, in most mythologies the formation of the Earth and the existence of humanity are presented as the result of fights of the divine beings so that those who are overcome are the prime matter that is used for the several levels of material reality.

We must stress that at this stage nothing is said about the reason for the existence of matter (implicitly considered as primordial and eternal) and no indication is given of an evolution the Universe as a physical system. Rather, an initial chaotic stage is the first realty are deities that spring from that chaos are the source of order that makes the Earth fit for human life. Instead of a final state due to the changes imposed on matter by its activity according to innate laws, sometimes it is accepted that the world will go back to a chaotic condition, perhaps followed by a new re-structuring in an endless process due to a kind of irrational and blind "fate" that controls even the gods. It is possible to find this view in Indian mythology in Central America.

As a consequence of this way of thinking, science is impossible: there is no way to know nature in a way that — with theories and laws — will allow us to understand and predict its activity. Instead of science we have alchemy, divining by omens, magical spells, hidden forces... kind of knowledge that is privy to sages "initiated" in some group and who possess special powers. There is no objectivity or unanimity, but a variety of rites changing from one culture to another, from one city to the next one, from one sage or school to a rival one.

For the same reason, there is no special interest in physical measurements. It becomes necessary to have an elementary math for weights and lengths used in daily commercial activities for architecture, for establishing boundaries of agricultural fields and distances between cities, but there is no arithmetic or algebra as a subject of study. The symbols for numbers are ill suited for theoretical developments numbers appear mostly as records of royal taxes; of population census or commercial exchanges. We have an example of this utilitarian attitude in the Bible, where it is said — as a practical statement — that the length of a circumference is 3 times its diameter.

In Israel, where its religion is based upon a unique transcendental monotheism, nothing appears that would imply either a superstitious and magical attitude or an acceptance of divine attributes for nature and its elements. There can be no real god except Yahweh, thus avoiding any kind of possible rivalry or fight at the level of the divinity. In the more modern books of the Bible the concept of "creation" is explicitly introduced but matter is presented as devoid of intrinsic properties that would determine its way of acting: it is the decree of the Lord that the Sun should come up every morning and that the seasons should follow each other in the proper order for the good of mankind.

This is certainly a very valuable theological "model" centered upon a provident and just God wise and all-powerful, who does not act by arbitrary whims or any kind of necessity, but always acts moved by a loving concern He has made everything in an orderly way, and has imposed laws to his creation but they are laws imposed from outside (for instance, to make the seas remain within their boundaries). There is no real basis to develop science, but only a chance to discover more or less by sheer luck some useful healing properties in plants or other things. To study the heavens, except to determine a religious calendar and observance, was a danger it seemed to lean towards the pagan obsession with astrology. In different ways the entire culture of the ancient Hebrew people is only the description of its relationship to Yahweh at the historical, social and personal level without developing their own art (architectural or decorative) or any science. The "Wise Man" is the one who knows Law — the way Yahweh has established for Israel to remain faithful to God — who knows practical things and who is skilled in solving riddles or interpreting the saying of older sages. Nothing remains of the Wisdom of Solomon as his proper contribution to human culture: when the Queen of Saba visits him, she is impressed by his insightful answers to her multiple questions, but no systematic body of knowledge is mentioned.

Scientific–formal–geometrical models

The Greek cosmological model, from the first philosophers and geometers, appears in history as an effort to "save the phenomena", describing positions and motions that have no known cause, but no effort is made to even look for one. Geometry is the prototype of science, mostly because its inferences can be visually presented, while even a simple sum is practically impossible to do on paper if we use Roman numerals (Greek notation was similar). Perhaps it is hard for us to realize the importance of a proper symbolic notation when we try to do even the simplest arithmetic; a square root, all algebra and the following developments are unthinkable in the context of Roman and Greek ways of representing numbers.

Thus we find that concepts of order and relationships described by the geometry of regular figures and solids are applied to the Earth and to the movements of the heavenly bodies. Cosmology is considered adequate if, through combinations of geometric elements, it can explain the way the heavens look and foretell the motions of the planets, without giving reasons why things are the way we observe them.

By strictly geometrical arguments the Greek scientists before Aristotle could give a proof of the spherical shape of the Earth and, three centuries before Christ, Eratosthenes found its circumference and diameter. At about the same time Aristarchus measured the size and distance to the Moon, comparing it with the Earth's shadow during an eclipse and relating its diameter to its apparent size in the sky (half a degree). He even tried a correct method to find the distance to the Sun, but without being able to measure angles with the required precision. Aristarchus also proposed that the position and apparent motions of planets in the sky against the background of stars could be described in simpler terms supposing the Sun as their center it was not necessary to wait until Copernicus.

But the same geometrical reasoning led to think that the Earth did not move: an orbit tens of millions of kilometers in diameter (at least 20 millions according to Aristarchus) seemed to require that during the year some stars would show a different position against the background of the more distant ones: this could not be observed, thus leading to the idea that the observer on Earth was not moving. This basic reason was behind Ptolemy's system, even if it required the introduction of cycles and epicycles as a way to calculate the position of the planets just at the minimal level of naked eye observations.

A clear indication of the purely formal value of the proposed astronomical constructions is found in the concept of an orbital motion about an imaginary center, where no physical object is found. The planet must move around an abstract center (of an epicycle), that also moves in a circle around the Earth. Further developments ended by making the center of this main circle fall also in empty space what we observe is the result of combining circles upon circles in a way that resembles how in modern times a graphical system can produce straight lines, triangles and rectangles, by combining circular motions. We can also represent a complex wave of any sound by the sum of sinusoidal curves of the correct amplitude and frequency.

Therefore it was not considered important to decide if the Earth and planets move around the Sun or if the Earth is motionless: there was no physical reason why one or the other should take place. The "center" was not a source of any force, but just a geometric point where we placed a compass to draw orbits for which no reason was known, and it was not even sought. With a purely abstract Physics (truly just a set of philosophical presuppositions) the structure of the Cosmos was described in terms of "elements" endowed with properties and positions that give only the qualitative impression of order. Four elements — earth, water, air and fire — constitute the material world at the earthly level, the environment where we find imperfection, change and decay. A fifth element — "quintessence" — constitutes the heavenly bodies, in a realm where immutable perfection requires strictly circular movements at constant speed and spherical polished bodies, that eternally shine because that is proper to their nature, of a dignity above that of any matter below the Moon.

We find the idea of "natural place" as the basic reason advanced to explain the hierarchical ordering of things: higher dignity requires a higher place. The Earth, the heavy and most obviously inert and crass element, must be at the bottom. The heavier a body is, the more it seems to seek the lowest possible point, from all directions, not because the Earth attracts it, but because that is where its natural tendency requires it to go. Against a current actual misconception, the position of our planet is not one of honor, but just the opposite: it occupies the center as the lowest point to which all falling bodies must go if left to themselves.

The geometrical model was the framework for the development of Astronomy almost until Newton. The Alphonsine

Tables of Alphonse X of Castile, the models of Tycho Brahe, Copernicus and Kepler, are varied methods of describing the data with abstract constructions, more or less complicated, but all without physical reasons to support them. Perhaps Kepler, suggesting something like a magnetic influence of the Sun upon the planets, begins to point to interactions or forces in the modern sense. But his geometric obsession is still evident when he tried to relate the distances of the five known planets to the Sun to the five regular solid bodies of Euclidean geometry. He had to finally accept that the orbits were not circles, but ellipses. But the "distorted circle" was an ugly fact that he never quite felt comfortable with. Geometry was an almost poetic way of describing a world that the Creator had made with order and measurement, according to a biblical expression. The Greeks could also assert "The Divinity is a geometer".

The historians of science and philosophy can describe in detail how cultural trends may be related to the development of the sciences that we now consider as part of our modern world. We will simply mention here that there was a wide-spread acceptance of the idea of "influences" of the heavenly bodies upon the Earth and human life, due to some kind of unspecified "pre-established harmony": the positions of the planets determined the growth of plants and animals and even the formation of metals within the Earth, and a comet could announce or cause pestilences and other calamities.

The entire Universe was conceived as a tissue of real relationships, not understood or quantified, that were due to the Creator (in the Christian way of thinking) in order to help human life. In an empirical way, science develops toward Chemistry instead of the esoteric ideas of alchemy, and towards a Physics that is, first of all, Mechanics. The planets are studied more and more as material bodies, without any astrological connotations, even if they are beautiful in the sky and they serve to establish a suitable calendar (the week reflects the seven bodies that are observed moving against the background of the "fixed stars"). Thus the way is laid to find a new model: the Universe as a perfect "machine", even if the meaning of this word will evolve with time, but always keeping the basic idea of quantifiable interactions rooted in the nature of matter itself.

In the meantime, the Indian symbols for numbers and the positional notation in a decimal system are introduced in Europe by the Arabs. Later on we find the zero, the decimal point, symbols for addition, subtraction, multiplication, square root, the equal sign: steps that render computation faster and clearer, and that will lead to a simple way of detecting quantitative relationships between numerical data obtained from experiments.

Models based on scientific causality

(a) *Mechanical model*

A new way of thinking about the Universe slowly grows from philosophical and scientific ideas around the 17th century. Instead of considering as the main property of matter the concept of "proper or natural place" — that should determine the position and movements of the different elements, even in the sky — we see the progressive acceptance of a modern viewpoint: matter is the same at all levels, and its activities and structures are due to "forces" intrinsic to it. Their reach and way of acting can be experimentally determined and mathematically described, with equations that are more flexible and meaningful than the geometrical drawings of previous times.

Galileo's work on moving bodies stresses that the proper scientific method is the quantitative experiment, something rarely present in the study of "Natural Philosophy", even if we do find some important instances in the middle Ages and even in Aristotle. It is the beginning of Mechanics, and it leads to a radical new idea: a body has no predetermined tendency to be at rest or in uniform motion. Thus we get the concept of inertia, together with that of friction, both used to explain the continuation of a motion once started and also the obvious fact that finally the motion stops. The accelerated motion of falling bodies establishes a mathematical relationship between space and time (Galileo: in free fall, the distance is proportional to the square of the time) even if the lack of an accurate clock makes exact measurements almost impossible.

The moment has arrived when a description of the Universe in clearly physical terms becomes possible, thanks to Newton's genius. His three laws of Mechanics establish a basis for Physics as important for the new science as Euclid's works had been for Geometry. The world is understood as a machine and its driving force is Gravity: for the first time an interaction is proposed where a body acts upon another (something suspected in magnetic effects) without contact. This is so daring and surprising that Newton himself didn't quite affirm it: his prudent wording was that "everything happens as if bodies attract each other with a force that is proportional to their masses and inversely proportional to the square of their distance".

The laws of gravitation and mechanics give a perfect explanation for planetary motions and the orbits they follow: the solar system is a magnificent "clock" of amazing perfection. The fall of a body towards the center of the Earth does not obey any natural "tendency" to occupy a place dictated by its dignity, but rather follows from the same force that keeps planets and satellites in their orbits. Still, Newton has to face some problems from Physics and Philosophy that he cannot solve.

Gravitation seems to act instantaneously, and at a distance. Newton proposed the existence of absolute space as an eternal container for bodies, but without any physical property. Such space should be infinite (without limiting borders or center) to avoid a gravitational collapse of all masses to the center. Time should also be infinite, since a beginning of time implies a previous time as well. Since Newton did accept the fact of creation from the Bible, he ended by identifying space and time with divine attributes, something that is not a concern of Physics. And to keep the planetary system from falling into chaotic disorder — due to perturbations of all bodies interacting — he accepted that a divine intervention was needed from time to time to maintain stable orbits.

Even with those limitations, Newton's work changed forever the scientific panorama. Laplace suggested gravitational condensation of a spinning cloud to explain the formation of the planetary system. Perturbations in the orbit of Uranus were sufficient to infer the existence of Neptune, and it was found where the math of Adams and Le Verrier had predicted its position. From reports of previous sightings, Halley for the first time correctly predicted the return of a comet, now named after him. The mechanical description was the most perfect example of order and certainty, and it seemed that all science should end up by explaining everything in terms of its laws, a dream that was realized to a great extent when sound, heat and even light seemed reducible to vibratory motions, basically similar to the way matter acts at the macroscopic level.

Gravitational theory allows us to derive and give a reason for Kepler's Laws. It appeared possible to attain the dream of reducing Physics to Newtonian mechanics, where the concept of "force" is central: all forces, gravitational or mechanical, will produce accelerations even if other effects — due to other forces — have to be accepted to explain changes in the states and properties of matter. This view reaches our time when matter itself is defined in terms of its four forces or interactions: gravitational, electromagnetic (long range interactions), strong and weak nuclear (extremely short range).

The science of optics becomes fundamental for the study of the information contained in light received from the stars.

We do not see because our eyes send any kind of ray towards the object (a common opinion until 1025) but because "something" material is sent by the star towards the observer. Newton thought that it was a stream of particles, but that theory was improved by Huygens, speaking of "waves" that end up by being electromagnetic, not mechanical. The colour of light will indicate the temperature of its source, and spectral analysis will be the key to find out the composition and motions of the stars.

Indications of evolution are also found in the areas of Geology and Biology, with cogent reasons to think in terms of enormous time spans for the age of rocks and for life forms that no longer exist. Fossils — previously considered as mere effects of a playful nature — are now seen as petrified remnants of real living, but now extinct, animals and plants. It seems that we should accept the ancient saying of Greek philosophers: panta rei, everything changes, a concept that is finally overpowering when — just about a century ago — we came to realize that the Universe itself had a sudden violent beginning from which it has evolved to its present state. But that insight required an unsuspected rethinking of many "certainties", a new viewpoint that led to a more drastic revolution than any previous one.

At the same time that theoretical Physics was expanding its reach, the industrial revolution extolled the machine, constantly improved and diversified, giving scientists the necessary means to get more and more precise data in all their experiments. With that higher precision applied to angular measurements with better telescopes, it was finally possible (in 1838) to prove stellar parallax and to determine stellar distances, opening to astronomical study the universe outside the solar system. In the 20th century the structure of the Milky Way could be established, with our position within it, and we can now speak of innumerable other galaxies and of the way the whole Universe appears from our viewpoint.

One can sense the optimistic feeling that was common among scientists of the 18th and 19th centuries, frequently accompanied by an air of satisfied superiority with respect to other fields of intellectual work that cannot rely upon experimentation and measurement. The hero of this new world is Newton, and it seemed that all that remained to do was to improve the accuracy of the data ("get one more decimal figure") and apply that methodology to all levels of nature, from atoms and molecules to the vastness of space.

(b) *A dual model: physical and geometric (Relativity) and quantum mechanical*

The Newtonian approach, seeking measurable activities and a suitable mathematical formulation, continued to give important fruits that we can only mention in general terms. The discovery of magnetic and electrical forces, the synthesis of electromagnetism in Maxwell's equations and their application to the nature of light, the negative outcome of the Michelson–Morley experiments (Cleveland, 1887) and the new understanding of the atom from natural radioactivity, are the most important steps towards a more complete description of physical reality. But the new model is framed during a time of unbelievable change of ideas, centered upon the Theory of Relativity at the astronomical level, and Quantum Mechanics in the infinitesimal world of atoms, in the first half of the 20th century.

In 1916 Relativistic Cosmology, developed from Einstein's genial insights, explained gravitational interactions by accepting geometrical distortions of empty space, reacting to the presence of masses and determining orbits as the paths that twisted space imposes upon moving planets. For Newton, space was just a container, empty, inert, eternal and infinite, but now it appears as an active component of the material world, with a changing geometry — that requires going beyond Euclid and common sense — that implies a fourth spatial dimension, so that the observed tri-dimensional reality can be distorted towards it. Time is also included as a physical parameter in the description of the structure and activity of matter: processes by which we can detect its flow occur at a different rate under the influence of gravitational or mechanical accelerations. We have to consider the whole universe as a physical system with manifold influences, with a global geometry and a necessary evolution of its tri-dimensional volume either expanding or contracting.

Those who delve deeply into General Relativity and the Cosmology based upon it feel very frequently that its conceptual simplicity and radical newness makes them admire its beauty even more than its mathematical brilliancy. It is said that when asked what his reaction would have been if the experimental tests were contrary to his theory, Einstein replied that he would have been saddened because the Creator had wasted the opportunity to do something intellectually so beautiful. It is the opposite of the remark attributed to Alphonse X when presented with the complex theory of cycles and epicycles: "I think that, if I had been present at the creation of the world, I could have suggested something simpler". Due to its compelling logic, even against the tenets of our common sense, the relativistic cosmology and the geometric nature of gravity were almost universally accepted with enthusiasm by the best scientists.

Einstein's model suggested, for the first time within scientific circles, that the universe is finite but unlimited, and that it evolved from an initial state — at a time in the past that we might be able to determine — to a future condition imposed by the play of physical forces (work of Friedman and Lemaître). A viewpoint that appeared so unexpected and daring, that Einstein himself found it instinctively repelling and incompatible with his ideas or science, even if he later surrendered to experimental proofs.

The central idea of the new Cosmology might be found in the concept of multiple and universal interactions carried to its logical consequences: at all levels, it describes activities of a unique reality that shows itself in multiple ways. Explicitly, it involves the equivalence of mass and energy, so that "matter" includes particles and waves, the physical vacuum and even space and time.

While Relativity dealt with the totality of the Universe, Quantum Mechanics — to which Einstein also contributed, together with Max Planck, Bohr, De Broglie, Heisenberg, Pauli, Dirac and Schrödinger, mentioning only the best known — presented us with an intimate description of matter impossible to reconcile with our intuitions. The development of experimental physics, of the atom and its basic elements, leads to a synthesis of physical activity as due only to the four forces previously mentioned, that have different intensities, ranges and effects upon "elementary" particles that appear more and more complex and which cannot be described in terms of everyday language.

The nature of light — particles or waves? — is stated without a clear answer: it is "something" that can manifest itself as a particle or a wave, depending upon the experiment we perform. The same answer must be given regarding the more "material" components of ordinary bodies, atoms and molecules, which can behave as waves. We cannot distinguish the effect of gravitation from that of a mechanical acceleration in a spaceship, and we certainly are at a loss when told that energy (always considered as something "immaterial", an accident and not a substance) can be changed into particles, and also the opposite.

At present we still lack a unifying synthesis between the viewpoint of continuous fields of Relativity and the discontinuity of energy — and even the space-time substrate — in Quantum Mechanics. The quest for unity reaches what one could consider a desperate level when proposing multiple

Universes (clearly defined as undetectable and thus outside scientific methodology): we are presented with a new "mythology" of hypothetical constructions where hints from particle physics are expected to support the new cosmological hypotheses, that, in turn, must provide a reason to expect the existence and properties of those particles.

But — after many years — there are no data to lead us to accept those ideas and experimental verification appears impossible even in principle: we seem to go back to the way of thinking of ancient Greece when epicycles and sidereal substances provided a merely formal explanation of things that were not understood We now have, instead, multiple dimensions and mysterious types of matter or energy.

We thus arrive to the present, when we see no clear path towards a cosmology that will truly explain the world, at all levels, from a single viewpoint. We are forced to work with a methodology based upon practical constraints and logically coherent but without strict proof. It is expressed in the so-called "Cosmological Principle": the Universe is homogeneous and isotropic. There are no peculiar directions or places when we examine sufficiently large volumes. Therefore, every observer anywhere would see the same structures in those surroundings and would attain a similar description of the world.

This confidence arises from the mechanical model carried to its limits, and it permits us to obtain far reaching consequences about the structure and evolution of the Universe. It rests upon many observations and experiments that are understood within a framework of general statements: large volumes of the universe in opposite directions show the same variety and abundance of objects and structures (homogeneity); the spectral analysts of light from stars and galaxies indicate the same elements ruled by the same forces that determine wavelengths and energies proper to each element (universality and constancy of laws and material properties). Matter is the same at all places and we can extrapolate our laboratory measurements to the same elements everywhere under the same physical conditions. Physical laws are universal and unchanging. They can be applied to infer the past and to predict the future at any place and time, if we know the initial conditions and the laws of each system. This will be true, of course, within the margin of error of our measurements.

The current success of astrophysics can only be explained in the context of this cosmological principle as an objective description of the observable Universe.

The subject and limits of science

In 1948, the great telescope of Mount Palomar was officially dedicated, and for many years remained as the largest and most productive in the world. It was then said that "if the Universe that modern Science shows is truly admirable, more admirable still is the power of the human mind that has discovered and analyzed so many marvels". Scientific work is not finished in any field, each new experimental datum or observation gives rise to new questions. It has been said also that "Science is born by answering questions and it develops by questioning answers".

Science, by its methodology, can only speak of whatever can be measured and that can be checked in an experiment that anybody having the necessary means can perform. But prestigious scientists are aware of the limitations of that methodology that leaves outside the scientific treatment; everything we call the Humanities and that embraces the most valuable aspects of human activity: the Arts, Ethics (human dignity rights and duties), Literature, family and social relationships. It cannot deal even with the basic desire to know what produces Science with its never satisfied quest to find Truth, Beauty and Goodness, things that cannot be measured or introduced into a mathematical equation.

Abstract thought and free will cannot be described as properties of matter or even be explained by the play of the four forces that define it. Even when dealing with the material Universe the most basic question (according to J.A. Wheeler) is "Why is there something instead of nothing?" The next one, he says, is "What relationship exists between human existence and the properties of the Universe at the first moment?" And he goes on to admit that if we have no answer, we should confess that we don't understand anything completely. But the final "why" and "what for" cannot be expressed mathematically or as the result of a measurement.

These limitations of the scientific method are unavoidable and they cannot be attributed to it as temporary limitations. On the other hand, they should not be considered a reason to undervalue or despise the sciences. Every type of human knowledge is partial, even if true, and thinkers from all times and cultures can contribute to the common treasure of human wisdom: there is no danger that we might exhaust the field of what can be known and understood. From St. Augustine to Newton and Einstein, the clearest minds have marveled at their own discoveries, while confessing the immensity of the boundless ocean they sensed lay behind their achievements. Only those who know very little can think that they know it all.

There is a story — for whose accuracy I can't vouch — regarding a letter from the man in charge of the Patent Office in the U.S towards the end of the 19th century, addressed to the Government that, as usual, found itself short of funds. The letter proposed closing the Office, "because everything had already been invented". And a well known scientist, towards 1925, tried to discourage a student who was planning to study university Physics: he should get into another field, "because there was physics left only for four or five years". There is still no danger of that.

Nature, and Nature's laws lay hid in night. God said, "Let Newton be!... and all was light".

Alexander Pope (1688–1744), British satirical poet

"I do not know what I may appear to the world, but to myself I seem to have been like a boy playing on the seashore, and diverting myself in — now and then — finding a smoother pebble or a prettier shell than ordinary, whilst the great ocean of truth lay all undiscovered before me".

Sir Isaac Newton, English mathematician and physicist, c. 1700

"One thing I have learned in a long life: that all our science, measured against reality, is primitive and childlike — and yet it is the most precious thing we have".

"You imagine that I look back on my life's work with calm satisfaction. But from nearby it looks quite different. There is not a single concept of which I am convinced that it will stand firm, and I am uncertain whether I am in general on the right track... I don't want to be right... I only want to know whether I am right".

Albert Einstein

TIMELINE: SEVEN CENTURIES

A CALENDAR OF SELECTED DATES RELATED TO SCIENCE

1054 Chinese observers describe a Supernova in Taurus (now visible as the Crab nebula).

1500 Leonardo da Vinci performs anatomical dissections.

1504 Henlein makes the first pocket watch.

1514 First version of Copernicus' model — Vesalius publishes "Anatomia Humana".

1540 Rheticus publishes a summary of Copernicus' theory.

1543 Full printing of Copernicus' work, after his death.

1550 Geronimus Cardano proposes an evolutionary theory.

1572 Tycho Brahe observes and describes a Supernova.

1574 Tycho Brahe opens the first permanent observatory in the island of Hven.

1577 Tycho Brahe describes a comet, placing it three times farther than the Moon.

1578 Mexico University opens a Medical Department.

1582 Gregorian Calendar introduced (based on work by Fr. Clavius, S.J.).

1586 Stevinus shows that objects of different weights fall with the same speed.

1590 Galileo publishes his mechanical experiments in "De Motu".

Janssen builds a compound microscope.

1600   Gilbert describes the Earth as a magnet controlling the compass.

1604   Kepler observes a Supernova, described in a book in 1606

Galileo: In free fall, distance is proportional to the square of the time.

1605   Francis Bacon: Observation is the way to advance in science.

1608   The terrestrial telescope is in public use.

1609   Galileo uses the telescope in astronomy.

Kepler publishes "Astronomia Nova" with two laws of planetary motion.

1611   Sunspots observed by Galileo, Harriot, Fabritius and S.J. Scheiner.

1617   Briggs introduces logarithms of base 10.

1619   Kepler publishes the third law of planetary motion in "Harmonice Mundi".

1620   "Novum Organum" by Francis Bacon: The coasts of America and Africa fit 1624.

1624   Pierre Gassendi measures the speed of sound in air.

1633   Trial of Galileo in Rome.

1637   Descartes introduces Analytic Geometry.

1645   Von Guericke invents the air pump to produce a vacuum.

1656   Huygens discovers that Saturn has rings.

1657   Robert Hooke demonstrates that all bodies fall equally fast in a vacuum.

1665 Newton develops his ideas on calculus, gravitation and optics.

1668 Newton makes the first practical reflecting telescope (Newtonian).

1670 For the first time clocks are made with minute hands.

1675 Römer estimates the speed of light with a 25% error.

1678 Huygens explains light as a wave (published in 1690).

1687 Newton's *Philosophiae Naturalis Principia Mathematics* is published.

1728 James Bradley discovers aberration of starlight, proving Earth's motion.

1744 Lomonosov correctly interprets heat as a form of motion.

1758 As predicted by Edmund Halley in 1705, Halley's comet returns.

Mathematical symbols[2]

876 The zero is used in India.

1202 Fibonacci introduces in Europe the Arabic (Indian) numerals.

1492 Francesco Pellos introduces the decimal point.

1514 First use of + and − signs.

1525 Christoff Rudolph introduces the square root symbol.

1557 Robert Recorde uses = sign.

1631 William Oughtred introduces × sign for multiplication.

Thomas Harriot uses raised dot and > and <.

1659 Symbol ÷ for division introduced by Johann H.

References

1. G. Ochoa and M. Corey, *The Timeline Book of Science* (The Stonesong Press Inc., NY, 1995)
2. J. Carey, (ed.) *Eyewitness to Science* (Harvard University Press: Cambridge, MA, 1995).

This section has been adapted from previously published book by the author in Julio A. Gonzalo and Manuel M. Carreira, *Everything Coming Out of Nothing* (Bentham Science, 2012).

Galileo Galilei (15 February 1564–8 January 1642)

3. Frank Sherwood Taylor: The Man Who was Converted by Galileo

John Beaumont

The Galileo case has become an important part of the mythology of science. The standard account until relatively recently has been that a great revolutionary genius was scandalously treated by the Catholic Church, even to the extent of being tortured for his beliefs. The case has been seen as a triumph of reason and rationality over an obdurate and bigoted institution, still stuck in the supposed ignorance of the Middle Ages.

Well, now that the great work of Pierre Duhem[1] has been brought into the light, that view of the Middle Ages is in any case completely refuted. In addition, modern work by great historians of science has restored some balance to the debate around Galileo. The episode may not have been the high point in the Church's relationship with science, a relationship that has been a very positive one on the whole, but neither is it now seen as an attack on science by the Church.

But there is more. Far from the Galileo case being a reason for turning one's back on the Catholic Church, it has in at least one case played a major part in the conversion to the Catholic faith on the part of an eminent scientist. The person in question is Frank Sherwood Taylor.

The Background[2]

Frank Sherwood Taylor was born at Bickley in the county of Kent in England on 26th November 1897. He was educated at Sherborne School, where he won a classics scholarship to Lincoln College, Oxford. However, World War I intervened and Sherwood Taylor found himself on the Western Front. At Passchendaele, he volunteered to take place of an older man in a front-line action. He was seriously wounded, and lay on the battlefield for several hours before being found. Over the next nine months, he underwent fourteen operations. He recovered eventually, though always subsequently walking with a limp. In 1919, he took up his place at Oxford, but was allowed to study chemistry, a subject that had always fascinated him, rather than classics.

Oxford University has always had a fine reputation for chemistry and Sherwood Taylor benefited from the tuition of Oxford's best in this field. In 1921 he obtained his degree with distinction (degrees for war service candidates, which were shorter in length, were not given classes). From this base, Sherwood Taylor embarked on an initial career as a chemistry teacher in a number of schools. During this time, his other interest in classics set him on investigating Greek alchemy. He enrolled as a part-time student at University College, London. A preliminary thesis won him the Oxford postgraduate degree of BSc in 1925. The full thesis, "A Conspectus of Greek Alchemy", earned him his London PhD in 1931.

As a school teacher, he produced a number of school and university text books on chemistry and in due course moved on to university teaching. This was in 1933 when he was appointed Assistant Lecturer in Inorganic Chemistry at East London College (later to be Queen Mary College), London University. At this point, he did some research work, but his most notable activity was to write the first of several books on general science for the average reader, *The World of Science* (1936).

The Early Religious View

So, what was Sherwood Taylor's religious background? His father was a solicitor, referred to by his son as a "non-militant Huxleyan agnostic". His mother was a good Anglican. He was baptized and brought up in the manner of most children of his social class at that time. He describes this vividly:

"My religious education was, I fear, almost worthless for one who was destined to emerge into a non-religious society. I attended Matins on Sunday with my parents, I said my prayers and I must have read and been told some Bible-stories... I will not question too much the formal mode of instruction that I received in early childhood: rather would I lay stress on the fact that I did not see the Christian religion manifested. All the family, except, perhaps, my father, regarded themselves as Christians, but our conduct was not regulated with reference to God, but to the social rules which prevailed in the early Edwardian period. The most rigid was that there should be no mixture of classes: whatever might happen in the country, it was imperative in the suburbs that ladies' children shouldn't speak to poor boys...

Prayer was for me like cleaning my teeth, to be done irrationally before bed-time: church was the way women and children spent Sunday morning, while the men — the admirable part of creation — played golf. Nobody prayed in church on weekdays... In any event it is clear that unless children see their parents and fellows openly acting with reference to God, nothing that they are told about religion will be of any use".[3]

This sort of upbringing was undoubtedly typical for one with his social background at that time in England, with its established religion. It was no surprise that Sherwood Taylor would react against it when he discovered science:

"I had been reading odd bits of popular science since I was eleven but when I reached the age of fifteen, I encountered systematized science, which presented a world of logical precision, certainty and order, which gave me the strongest aesthetic pleasure".[4]

At this point in his life, Sherwood Taylor was prepared to accept the naïve and rather simplistic view supporting mechanism, materialism, and positivism. But even then he had some doubts:

"What, then, did I find that could not fit into my mechanical philosophy? First of all my self, which I knew to be alive and thinking. I was not content to be a positivist in this matter... I could not see how by adding atom to atom we could get life and thought".[5]

His doubts regarding the scientific view of mind extended to the scientific view of matter:

"I adopted evolutionary theory as the scheme into which I fitted my ideas of the cosmos, yet found it not wholly satisfactory. There seemed a great dearth of intermediate forms at the significant stages and the whole of nature seemed much too good for its job. Why on Earth should a plant produce a wild rose for a business-like brute of a bee who simply wants a pub-sign, the Pollen and Nectar — so to speak? It did not seem to matter whether the beauty was in my eye or in the rose — what place had it in the evolutionary scheme? I could understand seeing beauty in a woman as an inducement to augment the race, but where was the link between women and wild roses? If they and I were designed by the source of Beauty — yes".[6]

Galileo Comes on the Scene

And then Galileo came to the rescue! In essence what Sherwood Taylor found was that, as he put it, "the case has been disgracefully distorted by the opponents of the Church".[7]

Sherwood Taylor recounted the story on a number of occasions, most expansively as follows:

> "At this time... I found by the oddest of providences, my first approach to the Catholic Church, The Rationalist Press Association wrote to some other person of my name asking him to lecture: the letter was sent to me in error and I offered my services. What, I wondered, would they like to hear about? What was the greatest crisis for rationalism in the history of science — my specialty? Surely the case of Galileo, of whom I then knew little; I would get up the subject and give them the lecture. I did so, and I went on to write a book about his life. As I studied the documents and detailed histories, I became aware that the usually accepted Galileo-legend was full of deliberate distortions by anti-Catholic and so-called rationalist writers.[8] The Catholic Church did not, it was true, play a very admirable part therein, but it was quite clear that she had been wickedly traduced".[9]

The book that Sherwood Taylor wrote after the original lecture was *Galileo and the Freedom of Thought*, published in 1938. The first impression that one gets from reading it is how understated it is in respect of what later happened to the author. Yes, Galileo was the focus for his eventual conversion to the Catholic Church, but there was much thinking to be done and issues to be considered before the final step was taken and he was received into the Church. This took place on 15th November 1941.

What was it specifically, then, about the Galileo's case that prompted Sherwood Taylor to change his mind and look in a different direction? There are a number of points that he raised and developed.

Firstly, and contrary to what is often claimed, Galileo did not demonstrate that the earth revolved around the sun.

Galileo thought he had proofs, of course, the main one relating to his theory of the tides, but his expression of this in the *Dialogue Concerning the Two Chief World Systems* was shown within a year to be subject to serious mathematical error. Other proposed proofs were his telescopic findings of mountains on the Moon, its rugged surface, the plethora of stars everywhere, the four bodies revolving around Jupiter, and phases in the appearance of Venus. These, though supportive of the Copernican theory, did not provide a physical proof of it. What they did was to demolish geo-centrism and the Aristotelian cosmology.

In addition, at this time (which was before Newton's discoveries, of course) there were relevant objections put forward to the motion of the earth (even before Galileo it was known that the orbital speed was great), as for example the questions why there were no huge winds blowing all the time across the globe; why falling objects didn't fall backwards; and why everything movable didn't fall off the face of the earth?

Faced with this situation, the Church quite reasonably required Galileo to state his views as a hypothesis only. It was long after Galileo that science mustered really convincing proofs. One was Bessel's observation of stellar parallax in 1837, with much improved equipment. Some of Galileo's critics during his lifetime were right in pointing out that the parallax of the nearby stars against the background of the more distant stars was not at all observed (due to the primitive telescopes available at that time).

The other convincing proof was the pendulum experiment, performed by Foucault in 1851. Less significant were two earlier proofs. They were connected with Roemer's deduction in 1675 of the speed of light from the observation of the motion of Jupiter's moons and with Bradley's observation in 1728 of the aberration of light.

The second matter that Sherwood Taylor came to appreciate was that although the Roman Curia, in its approach to the case, was entering upon shaky scientific grounds, everybody at the time knew that no Papal infallibility was involved in the case:

> "The Church claims to be infallible only in matters of faith and morals intentionally promulgated as articles of faith by a Pope or General Council. The decision of 1616 was only an opinion of a committee of experts concerning what could safely be believed: it was not infallible, and has in fact been reversed".[9]

Thirdly, the Galileo case then (as opposed to the Galileo legend) is a much more nuanced matter than many people appreciate. One vital factor, which contributes to this, is the fact of a severe clash of personalities. Contemporary polemics among learned men at various universities, fuelled by Galileo's known strong temperament (Sherwood Taylor refers to his "sarcastic and witty pen"[11]), were already underway before the open conflict with the Holy See. In addition, the Protestant revolt made the case specially sensitive in Galileo's time.

> "At this period there was very acute controversy as to whether the Church was to be the interpreter of Scripture, as the Catholics urged, or whether the private individual was to be the interpreter, as the Protestants held".[10]

In relation to this latter issue, it is a great irony that in fact Luther and other Protestant leaders had already, in no uncertain terms, condemned Copernicus (who was feted in the Vatican gardens by a gathering of cardinals and bishops as he lectured there on his theory of helio-centrism). It is probable that at that time the Church was reluctant to contradict the Protestant leaders flatly. Many eminent scientists today are of the opinion that the Church authorities were simply prudent in the Galileo case.

Fourthly, there is the ultimate irony, that Sherwood Taylor appreciated, namely that while Galileo failed in his stated claim of presenting unassailable experimental proofs of the new theory, when it came to the interpretation of the Bible he bested the finest theologians of the Church (see his *Letter to the Grand Duchess Christina* (1615)). Galileo was right when quoting St. Augustine on the interpretation of the Bible. Both sides to the debate should have followed St. Augustine's wise advice:

> "We do not read in the Gospel that the Lord said: 'I send you the Holy Spirit to teach you how the sun and the moon go.' He wanted to make Christians, not mathematicians".[11]

Galileo himself wrote in the *Letter to the Grand Duchess Christina* that "the Bible was not written to teach us astronomy", and quoted the well-known statement attributed to Cardinal Baronius that "the intention of the Holy Ghost is to teach us how to go to Heaven, not how the heavens go".[12] Sherwood Taylor appreciated also how St. Augustine warned against reading hastily our own opinions into the Scriptures and fighting for them as if they were the teaching of the Bible. Sherwood Taylor summed up this aspect of the case in the following words:

> "It is clear that the Holy Office in 1616 acted foolishly, and that the texts in the Scriptures which seemed to imply that the earth was stationary and that the sun moved were obviously not intended to teach astronomy and were mere figures of speech. We use such figures today, when we speak of 'sunrise', knowing well enough that the sun does not rise, but remains stationary relative to the earth's orbit while the earth rotates on its axis".[13]

In respect of the approach often taken towards the Galileo case, little has changed since Sherwood Taylor's day. The insistence of the secularists that Galileo was shamefully ill-treated still persists. This is why Catholics need to arm themselves with the arguments of refutation. In this context, it is worth noting, as Sherwood Taylor himself did, that the sentence of Galileo to house arrest for the rest of his life was also not quite what it seems. He first lived in comfort in the houses of two friends, then retired to his own villa with a handsome pension from the Pope, where he continued his studies and was visited by scholars from all over the world. The tale often told that he was tortured is completely untrue, as is the claim that on being forced to recant he said, *sotto voce*, "But it does move" (*Eppur si muove*). This is just scientific hagiography. In addition, and this is a point not made by Sherwood Taylor, it seems pretty clear now that Galileo never dropped balls from the Tower of Pisa or from any other tower.[14] Finally, and what is the ultimate annoyance to the secularists, Galileo of course died as a good Catholic in 1642 and in the company of his daughter, a nun.

Another Argument

So, all of the above arguments show that there is indeed a good defence for the Church's decision in the Galileo case. Sherwood Taylor became aware of this and, as we have seen, set out several of the arguments. However, one extra point might be made here. It takes time to make these points and, unhappily, experience shows that in day to day life many will not listen. So, as Quentin de la Bedoyere wrote, in his always interesting *Second Sight* column in the main British Catholic newspaper, the *Catholic Herald*,[15] it is useful to have to hand another kind of argument. This is one that Sherwood Taylor did not make, since he was looking at the issues solely from his own position, but he would have appreciated it. It amounts to that very effective (rhetorical) tactic, the *tu quoque* ("You also") argument. As Bedoyere acknowledges, "it doesn't get you nearer to the truth but it can unsettle opponents, and make palpable hits".[16] So, if the Galileo case is brought up in discussion as an example of the Church's opposition to science, and time is limited, a more direct and effective response is to say, "That's a strange accusation to make given that the history of science is marked by its refusal to accept plain evidence, with far more serious consequences than Galileo".[17]

The point here is that there are, of course, many examples of cases where science itself did not follow the evidence. Take for example, as Bedoyere does, William Harvey's demonstration of the full circulation of blood. He published this in 1628 together with experiments and reasoned arguments. But the evidence was rejected by medical science and it was some further twenty years before it was accepted. Not exactly a glowing reference for the evidential objectivity of science one might think.

Bedoyere himself takes as a more powerful illustration to the Semmelweis case.

"More dramatic is Ignaz Semmelweis, the doctor who established, with indisputable results, that lack of hygiene led to a high rate of sepsis in midwifery and other medical treatments. This was in the 1840s. Despite the clear evidence, his view was rejected time and time again by the scientific community of doctors. It was not until the last decades of the nineteenth century that medical hygiene began to be more widely recognized. Meanwhile, a very large number of deaths, particularly in childbirth, had taken place. It is thought that the mental illness which preceded Semelweis's death was caused by the continued, tragic rejection of his proofs".[18]

As Bedoyere concludes, "The result of medical science's conservatism was the death of many thousands of innocent people. Galileo only got house arrest".[19]

Frank Sherwood Taylor's Positive Apologetics

Now back to Sherwood Taylor. The impact of the Galileo case was not to lead immediately to his conversion to the Catholic faith. What it did do was to open his mind to the possibilities in respect of this.

"I came to think that if the assertions of (the Catholic Church's) opposition to science were so ill-founded, so also might be all those stories of her wickedness, deceit and superstition which my Protestant and rationalist reading had put into my mind. I did not yet believe — but I now lay open to conviction".[20]

Sherwood Taylor was by no means the only person from the world of science to appreciate the Galileo case in this way. Darwin's so-called "bulldog", T. H. Huxley, referred to earlier, looked into the facts whilst in Italy and concluded that "the Pope and the College of Cardinals had on the whole the best of it".[21] Huxley, however, remained entrenched in his existing materialist philosophy, but Sherwood Taylor, albeit seven years after giving the original lecture, made the vital move. It is interesting to examine his positive reasons for becoming a Catholic. We shall come to the objective factors in a moment, but first of all it must be said, as so often is the case in the process of conversion, that there were personal elements that played a major role.

"I was fortunate enough to fall in quite separately with two Catholics of personal qualities which throw some light on what Christianity in action could be — and this, in my belief, is an essential in almost every conversion. If a man or woman goes about the world being charitable, humble yet inflexible in the faith, and radiating sanctity to those who have eyes to see it, the people who meet them will find their difficulties disappearing and will tumble one after another into the laver of regeneration. I suspect that one Living Christian is worth a shop full of hortatory treatises or an army of eloquent preachers".[22]

We do not know the names of these witnesses, but we do know that they brought to him also the witness of the Christian mystics (he had examined supposed thinkers of this kind from the non-Christian traditions, but found them much inferior).

Another vital preliminary came as a result of his writing a book under the title of *The Century of Science 1840–1940*:

"Like many scientists, I was woefully ignorant of modern social history and I expected to find in that century a smooth, easy triumphal progress. When I was driven to estimate the good and bad effects of science on human life in the years since the industrial revolution, I was brought to quite a different conclusion from that I had expected to find; namely that what had been and still is required by the world is not more knowledge, but better people; and that the scandalous economic cruelty and oppression of the nineteenth century was due to the wickedness of men rather than to ignorance about nature. I saw that as long as men wanted money and power for themselves more than they wanted happiness for others, oppression must continue...

I could not see any source of altruistic ethics other than a belief in God, nor could I see any other justification for following truth rather than lies, if the latter were more effective to achieve one's purpose".[23]

Both of these things, the witness of Catholics and an appreciation of the significance of good and evil, put Sherwood Taylor on the route to Christianity and the Catholic Church. His attitude by now was that "the pagan world was unsatisfactory; was the Christian world credible"?[24] This involved an examination of the relationship between his scientific credo and the Christian faith, basically in respect of two fundamental issues: (a) the existence of what may be referred to as the supernatural; and (b) the historicity of the events recounted in the scriptures.

In relation to the former, Sherwood Taylor was clear that "Science could neither affirm nor deny that which does not manifest itself in phenomena that can be treated by the scientific method".[25] And in relation to the essence of science, he goes on to emphasize the importance of the measurable to observable data. This gives us an immediate implicit reference to Fr. Stanley Jaki, the doughty champion of the limitation of science to what is empirical and measurable, to what is quantifiable. As Fr. Jaki expressed it on so many occasions, "Science, and by that I mean exact science, has quantities for its foundation. Indeed, it is nothing else than the quantitative study of the quantitative aspects of things in motion. Nothing more and nothing less than just quantities."[26] So, in the same way it was clear to Sherwood Taylor that although science was not inconsistent with the existence of the supernatural, yet that matter was actually moved by supernatural agency was not explicable by science. God could not be measured by the methods of science. But, if the mind was not reducible simply to matter, and as we have seen, Sherwood Taylor always saw them as distinct in a certain way, then if also the supernatural was analogous to mind, he must regard the supernatural and therefore God as entirely possible. It was from this somewhat agnostic position that he moved forward, though in later life he did write in great detail, both about the relationship of science and religion and the question of the proof of the existence of God.[27]

What about the question of historicity? Once Sherwood Taylor became convinced of the limitations of the scientific method, this question became one on which his mind became gradually easier. He understood that the events of the past were no longer observable. There was a radical difference between the kind of evidence on which one asserted that salt crystallizes in cubes, and the evidence on which one might assert the proposition that Christ did or did not rise from the dead. In relation to the latter, one thing could be said:

"The only recorded material was testimony, which could only be assessed as reliable or otherwise by a sort of judgment of value. The Gospels... read to me like the work of men that were telling the truth: no one could possibly have invented the character of Our Lord."[28]

Previously Sherwood Taylor had always set against the evidence of such testimony the uniformity of natural law. Now, he was able to see the possibility of the modification by God of the course of physical events. This recognition was buttressed by the influence of a particular book, the *Confessions* of St. Augustine. This text and its author had a major influence on him and show on his part an essential humility.

"Page by page it went home to me and I said to myself 'Mutatis mutandis, this fellow was in the same sort of trouble as mine. He has obviously a far keener intelligence than mine, far greater spiritual powers; why should not his solution be my solution?'"[29]

Sherwood Taylor remained for some time in a state of hesitation, unable to believe any longer in a mechanistic world, but unable to see the spiritual world in any terms other than a sort of hypothesis. On hindsight, he saw that "the gift of faith is a supernatural one and must be divinely infused."[30] His own case at the end involved a veritable Pauline road to Damascus, not usually associated with men of science.

"One day, suddenly and unexpectedly, I heard within me the words, 'Why are you wasting your life?' They carried instant conviction and removed all my difficulties. The world had changed: I knew what I was there for, I was endowed with goodwill, and it remained only to throw out the rubbish and rebuild."[31]

He at once sought out the Catholic Church and asked for instruction. It is not surprising, bearing in mind what has been said earlier, that he did not look to the middle-class public school Anglicanism of his youth. Here once more there is an emphasis on personal testimony. Sherwood Taylor was well aware of the strength of the formal arguments of Catholic apologetics, and had recourse to them, but again the major emphasis came from elsewhere.

"I will give some personal answers, the first and simplest and truest being that I had not a moment's doubt that it was God's will and intention to call me there. St. Augustine was my master, and again I had no doubt which church, if he were alive today, he would recognize as his own."[32]

Defending the Catholic Faith

As stated earlier, after his conversion Sherwood Taylor wrote in more detail concerning the relationship between science and religion and the question of possible proofs for the existence of God. He also wrote about the differences between materialism and Christianity. But, he went on to look at two other areas in particular. The first stemmed from his original discipline of chemistry, and here his work is still of major significance, the study of alchemy and early chemistry, out of which came *The Alchemists* (1951), for many years a leading work in its field.

More important in the present context, however, is the work in which he defended the role of the Catholic Church in relation to science. In addition to some general issues, referred to earlier, he set out very clearly an apologia for the Church's almost always very positive relationship with scientific issues. As early as 1944 he wrote an article entitled "The Church and Science"[33], which dealt with several issues, notably the interpretation of Scripture, Galileo, the theory of evolution, and general matter concerning the relationship between science and the Church. However, this aspect of his work was done most effectively in a pamphlet prepared for the *Catholic Truth Society* under the title of *The Attitude of the Church to Science* (1951), referred to earlier in relation to the Galileo case. Here we see, in addition to a more detailed treatment of the Galileo case than that in his story of his conversion, an examination of medieval science very much on similar lines to the later research of Fr. Stanley Jaki, the successor to some extent of Pierre Duhem. Sherwood Taylor here also takes up certain false tales of the Church's supposed opposition to science. Here can be found an account of particular such stories, including the Flat Earth story, the stories that the Church forbade dissection and chemistry, and the story that scientific men were burnt for their scientific opinions. In each case, he brings out the true position of the Church here, which is a world away from the allegations of the secularist camp. Lastly, Sherwood Taylor shows how certain admittedly unwise decisions made by the

Church in respect of Galileo have never been repeated. In this context, he looks at the way the Church has never intruded onto the field of science in respect of the theory of evolution. He would have seen at once the sound sense of a number of writers who have drawn attention to the differences between evolution as a science and evolution as an ideology.[34]

Conclusion

We can come to the Catholic Church from many directions. The secular world in which we live is increasingly strident in its opposition to any religious belief and claims that science leads away from belief in God. The example of Frank Sherwood Taylor shows that it does nothing of the sort. It also shows how an openness to the facts and a reasoned approach to this question can lead one, under God's grace, into communion with the one true Church.

References

1. See Stanley L. Jaki, *Uneasy Genius: The Life and Work of Pierre Duhem* (1984); Stanley L. Jaki, *Scientist and Catholic: Pierre Duhem* (1991).
2. For a relatively brief account of Sherwood Taylor's life, see *Dictionary of National Biography* (2004). For a more detailed account see A. V. Simcock, Alchemy and the World of Science: An Intellectual Biography of Frank Sherwood Taylor, *Ambix*, Vol. 34, Part 3, November 1987, p. 1.
3. Frank Sherwood Taylor, *Man and Matter: Essays Scientific and Christian* (1951), pp. 10–11.
4. *Ibid*, p. 12.
5. *Ibid*, p. 15.
6. *Ibid*, p. 16.

7. F. Sherwood Taylor, *The Attitude of the Church to Science* (1951).
8. *Man and Matter*, p. 16.
9. *Ibid*, p. 19.
10. *The Attitude of the Church to Science*, p. 12.
11. *Ibid*, p. 10.
12. *Ibid*, p. 11.
13. *Answer to Felix, a Manichean* (AD 404).
14. It is interesting to note that in his *Letter to the Grand Duchess Christina*, Galileo does not refer explicitly to Baronius, but refers to "a churchman who has been elevated to a very eminent position. The remark, which does not appear in the writings of Baronius (though it is unanimously attributed to him), was probably made by him in conversation with Galileo (we know that the two men, together with Copernicus, visited historical museums on occasions).
15. *The Attitude of the Church to Science*, p. 12.
16. See Lane Cooper, *Aristotle, Galileo and the Tower of Pisa* (1935); Stanley L. Jaki, *Galileo Lessons* (2001), pp. 1–8.
17. See Quentin de la Bedoyere, Enter the lion's den, *Catholic Herald*, 13th November 2009, p.9.
18. *Ibid.*
19. *Ibid.*
20. *Ibid.*
21. *Ibid.*
22. *Man and Matter*, pp. 19–20.
23. T. H. Huxley, *Letters and Diary* (1885) (Letter to Professor St. George Mivart, 12th November 1885).
24. *Man and Matter*, p. 20.
25. *Ibid*, pp. 20–21.
26. *Ibid*, p. 22.
27. *Ibid*, p. 23.
28. Stated in conversation with the author of this paper. For equivalent statements by him to the same effect, of which there are many, see for example, *Questions on Science and Religion*

(2004), pp.11-24; *Science and Religion: A Primer* (2004), pp. 4–5; *The Drama of Quantities* (2005) *passim*.

29. See *The Fourfold Vision: A Study of the Relations of Science and Religion* (1945); *Two Ways of Life: Christian and Materialist* (1947). In addition, the book referred to earlier, *Man and Matter* (1951) contains essays dealing with these themes.
30. *Man and Matter*, p. 25.
31. *Ibid*, p. 28.
32. *Ibid*.
33. *Ibid*, p. 29.
34. *Ibid*.
35. *The Month*, April 1944, p. 1.
36. See, for example, the following texts by Stanley L. Jaki: *Evolution for Believers* (2003); *Questions on Science and Religion* (2004), *passim*; *Darwin's Designs* (2006); "Evolution as Science and Ideology", *Lectures in the Vatican Gardens* (2009), p. 169.

Sir Isaac Newton (25 December 1642–20 March 1727)

4. The Limits of Science

Manuel Alfonseca

The fact that science has limits is an evidence. We do not know everything. The fact that science will always have limits, that there are things that we will never know, because they are out of our reach forever, is something that may be a surprise for some people. But it is true.

The limits of science do not always depend on our mental capacities. Sometimes the reason is deeper, and is founded on the structure of the universe. In such cases, the limits of science can be considered *theoretical* and therefore are unbreakable. In other cases, they are only *practical*, that is, in theory a problem may have an answer, but practical reasons make it impossible to attain.

In the following pages we will look at different sciences (mathematics, physics, biology and technology) and will focus on certain unsolvable problems. Sometimes the limits will be theoretical, sometimes they will be practical, sometimes both. As could have been predicted, in the cases of mathematics and physics most limits are theoretical. In the other two sciences, they are mostly practical.

The limits of mathematics

In the last decades of the nineteenth century, Friedrich Ludwig Gottlob Frege, a professor in the university of Vienna, undertook an ambitious goal: formalizing the whole of arithmetic in a set of axioms and a number of deduction rules, in such a way that every true theorem would be deductible from the axioms by a number of applications of the deduction rules. The result was a monumental book, *Grundgesetze der Arithmetike* (1983–1903), which introduced, among other things, a basic formalization of set theory and a cumbersome notation that was quickly replaced by Peano's.

Unfortunately for Frege, when the second volume of his book was about to be published, he received a letter from Bertrand Russell, proving that his formulation of set theory entails an inconsistent result. In Frege's set theory, there are sets that are not member of themselves (such as the set of all integers, which is not an integer), as well as other sets that are members of themselves (such as the set of all infinite sets, which is an infinite set). Russell then defined the following set: *the set of all sets that are not members of themselves*. It is easy to see that this set leads to a paradox: if it is a member of itself, it cannot be a member of itself, and vice versa. The paradox wreaked havoc with Frege's work, who had to add a hasty appendix to his book and later on abandoned his research on the fundamentals of mathematics.

Bertrand Russell, together with Alfred North Whitehead, attempted to complete Frege's work[1] in his monumental book *Principia mathematica* (1910–13). However, his work was destined to the same final defeat as Frege's. In 1931, Kurt Gödel proved his famous incompatibility theorem[2], which can be summarized thus: *every consistent formal system with a power similar to arithmetic is not complete (it contains undecidable true propositions)*. In other words, starting from a set of axioms and a number of deduction rules, either we will end up with an inconsistent result (as in Russell's paradox) or we will end up with true theorems that we are unable to prove.

The proof of Gödel's theorem reduces to the enumeration of a theorem that, informally, can be stated thus: *this theorem G cannot be proved starting from the axioms and rules of system S*. If this theorem is false, (i.e. it can be proved as indicated) it follows that system S is inconsistent (it proves a false theorem). Therefore, if S is consistent, the theorem must be true, which means that it cannot be proved in S (this is precisely what the theorem states).

Gödel's theorem was only the first of a family of incompleteness theorems that were proved in the following decades. The second was Alan Mathison Turing's theorem of the halting problem, which applies to a family of abstract computing devices he had designed, called in his honor Turing machines. These are universal computing devices, in the sense that every problem which can be solved by a digital computer can also be solved by a Turing machine. In fact, Turing machines are more powerful than digital computers, for they are supposed to have unlimited memory, which the latter do not have, although with current memory sizes this restriction can be ignored most of the time.

The halting problem can be enunciated as the solution to the following question: given a Turing machine (or any analogous mechanism) with a set of input data, will the machine stop, or will it keep on computing forever? For a given case, finding the answer may be immediate, but the problem is how to solve this question for every possible pairing of a machine and a set of input data. Alan Turing proved that the problem cannot always be solved, i.e. that the question is undecidable. In fact, assuming that the problem can always be solved leads to a contradiction, which means that it can not. Turing's theorem was proved to be equivalent to Gödel's theorem.

The fact that there are incomputable problems was a shock for computer scientists. This is a theoretical limitation for computing technology, very different from the practical limitation that makes *intrinsically difficult problems* impossible to solve. These problems are theoretically solvable, but to do it we would need a computer as large as the universe, working for more than the universe's life. For all practical purposes, therefore, intrinsically difficult problems are unsolvable. One of the best-known problems in this family is writing a problem that would be able to compute the perfect solution of a chess match, which would be unbeatable by any chess player. The program could in theory be written, but it would be impossibly slow.

During the nineteen sixties, American mathematician Gregory Chaitin proved another incompleteness theorem which can be stated thus: *the randomness of a set of integers is undecidable*[3.] This fact can be illustrated by the analysis of the digits of π, which fulfill all the tests of randomness currently devised by statisticians, but nonetheless clearly do not make a random series.

Chaitin's theorem entails that, given a series of integer numbers, their randomness can be impossible to prove. Therefore, artificial life individuals in a computer simulation would be unable to distinguish the pseudo-random series commonly used to simulate chance in genetic algorithms from real chance. It also means that, in our world, we also are, and always will be, unable to distinguish chance from providence.[4]

In 1989, Roger Penrose[5] proposed the following question, inspired by Gödel's theorem and its kin: *how can we know that a theorem is true, even though it is undecidable, i.e. it cannot be proved by mathematics?* This seems to point, in his opinion, to the consequence that human intelligence is qualitatively different from computational machines. We'll come back to this later.

The limits of physics

In the last centuries we have discovered that the weft of the universe can be tackled on different levels. While the next level has not been found, what happens in the preceding level cannot be explained, it can only be described. In consequence, for the lowest level known at any point, explanations are not possible, just descriptions. Let us look at history:

1. Eighteen century chemists discovered a large number of new substances. Since they were unaware of their composition, they could only describe them in catalogs of properties, but they could not explain those properties. In such a way, the famous book[6] published by the man considered to be the founder of modern chemistry (Anton Laurent Lavoisier) can be considered to be a catalog of the properties of the chemical species known at the time.

2. At the beginning of the nineteenth century (1803), John Dalton formulated the atomic theory, according to which molecules are made of atoms of one or more elements which combine with one another. Along the nineteenth century, while more and more chemical elements (types of atoms) were being discovered, many chemical properties could be explained, but the lowest level (atoms) could not be explained, it could only be described. Mendeleev's periodic table was just a catalog of atoms. A very intelligent catalog, but nobody knew why the different atoms had the properties they had, or why they grouped in such a way, rather than a different one.

3. At the beginning of the twentieth century (1911), after the discovery of radioactivity and the first elementary particles, Ernest Rutherford proposed a model of the structure of the atom, which goes down one level, the level of elementary particles (electrons, protons, neutrons since 1932), which grouping themselves to build atoms, explain their structure and properties. With this model, confirmed by the experiments of Henry Moseley and his discovery of the atomic number (the number of protons in the atomic nucleus), three levels were known: the levels of molecules, atoms and particles. Each level explained the upper one; the lowest (elementary particles) could only be described. Nobody knew, for instance, why

protons had a positive charge and neutrons none, it was just known that such was the case.

4. Along the twentieth century, in a similar process to what had happened in the nineteenth with the atoms, the number of elementary particles exploded: baryons, mesons, positrons, heavy electrons, neutrinos, photons... To put some order in the jumble, in 1964 Murray Gell-Mann proposed the quark theory, which goes an additional level down to explain the behavior of a family of elementary particles, the hadrons (which comprises baryons and mesons). With this theory, the charge of protons and neutrons is explained, as we now think that the former are made of two up quarks with a +2/3 charge, and one down quark with a –1/3 charge (2/3 + 2/3 – 1/3 = 1), while a neutron consists of two down quarks and one up quark (1/3 + 1/3 – 2/3 = 0). But nobody knows why quarks (and the other family of elementary particles, the leptons[7], which includes the electron) have the charge they have. We can only describe them.

We are just at this point (see table 1). We are currently aware of four levels: molecules, explained by atoms; atoms, explained by elementary particles; elementary particles, explained by fundamental particles (leptons, quarks and bosons[8]); and fundamental particles, which up to now have not been explained and can only be described.

Table 1. The fourth level of the structure of matter, current limit of particle physics

	I generation	II generation	III generation	Bosons
Quarks	u	c	t	y
	d	s	b	g
Leptons	v_e	v_μ	v_τ	Z^0
	e	μ	τ	$W^\pm$

What about the future? Perhaps one day we will be able to explain the behavior of the fundamental particles by discovering a fifth level, but then this fifth level (whatever it is) will not be explainable, it shall only be described. And so on and forth. The last level discovered will always be unexplainable until the next one is found.

The conclusion is evident: physical science will never be able to explain everything. This limit is theoretical, rather than practical, as there will always be a lowest level in our understanding of matter and physical nature. So the *theory of everything* that physicists are always talking about is something that will always remain out of our reach.

A further unsolved physical problem is the nature of time. The arrow of time (the fact that time is a directional dimension, which always goes from past to present to future) is strongly established by human experience and by the second principle of thermodynamics and the consequent irreversible processes (which can be found abundantly in many branches of science: mechanics, chemistry, radioactivity, quantum mechanics, the standard theory of particle physics...) However, in a glaring opposition to all these reasons, important scientists such as Einstein[9] deny the existence of time's arrow, just because in their pet theories time seems to be reversible in theory, thus providing us with a magnificent example of how even a scientific genius can forget that, in science, facts go before theories.

The limits of biology

As indicated before, the limits of biology are rather practical than theoretical. That is to say, some biological problems are so difficult, that it is very improbable that we will be able to solve them one day. Among these problems, we can select the following:

The origin of life. The possibility of replicating an experiment is one of the fundamental principles of scientific method. No discovery is considered valid until it has been confirmed by an independent scientist or team of scientists. When this is done, the result of the experiment becomes a part of the scientific heritage. As a consequence, an experiment or a fact can be considered scientific if it can be replicated.

The origin of life is a fact that probably happened only once during the whole history of the Earth. It is obviously something impossible to replicate. Therefore, it is not a scientific fact. What is it, then? A historic fact.

Historic facts are different from scientific facts, and are treated in a very different way: historians try to find documents that confirm that the fact did happen and describe how it happened. These documents are analyzed to estimate their credibility. In the case of the origin of life, which would be the documents? Fossil remains. But it is practically impossible that fossil remains of the origin of life can be found. Therefore, the origin of life will probably remain forever un unsolvable problem.

Is it not possible that one day we may succeed to create life in the laboratory? Perhaps, but this would not solve the problem, for we would never be sure that the process by means of which we would have achieved it was the same through which it happened spontaneously a few million years after the origin of the Earth.

The origin of cells. This is a related problem, but not necessarily identical, for not every biologist agrees that life started with cells. Most of them believe that some previous level of life occurred before any cells were there. Nucleic acids and proteins would have been subject to some kind of *chemical evolution* before a number of them got enclosed inside a cellular membrane and started to cooperate. In any case, the origin of cells would be a historic fact, similar to the origin of life. Fossil remains of the first cells are practically impossible to recover. Therefore, the origin of cells will probably remain forever an unsolvable problem.

Intelligent design or random evolution. Actually this is not a scientific problem, for it has no answer if we restrict ourselves to the scientific method. The existence of God (or His inexistence), the dilemma between the creation of the universe by an intelligent being or its spontaneous appearance out of nothing (whatever that is) is a problem that will never be solved by science. We will have more to say about this in a further chapter.

Brain, mind and conscience. The analysis of the human brain has advanced a lot in the last decades, but we seem to be no nearer to discovering what is conscience or self-awareness and how it appears. Neuroscientists usually start from the reductionist assumption that conscience is an epiphenomenon, something that emerges automatically from the work of billions neurons. This assumption has brought us nowhere. Perhaps we should start from a different assumption, perhaps self-awareness is not wholly reducible to a lower level. But then, it may be completely outside the scientific approach. We should remember here Penrose's suggestion that human intelligence is qualitatively different from computational machines, as mentioned above.

The problem of free will. This is related to the previous bullet and makes one of the two outstanding scientific-philosophical problems that modern science seems unable to focus (the other one is the arrow of time, mentioned above). Again, universal human experience confronts reductionist scientific theories, which come inescapably to the conclusion that free will is just an illusion. But there is an alternative explanation, according to which free will is an undeniable fact and materialism a wrong assumption. Taking into account the demise of deterministic materialism, which had its heyday in the nineteenth century and was ignominiously banned from science by three successive blows during the twentieth (Heisenberg uncertainty principle, chaos theory and quantum mechanics), current non-deterministic materialists should not be as sure of themselves as they seem to be.

The problem of artificial intelligence. Although this seems to be a technological problem, and thus should be considered in the next section, it is actually deeply related to the two previous bullets, which is why it is mentioned here. This question, however, will also be the subject of a further chapter.

Can disease and death be defeated? The art and science of Medicine is one of the most ancient in the history of mankind. Its main objective from the beginning was the fight against disease and death. The Hippocratic oath, traditionally taken by physicians, implicitly states this. However, up to now, and in spite of the tremendous advances of Medicine in the last two centuries, it has been a losing fight. Only one illness (smallpox) has been completely eradicated from the Earth. Many can now be cured successfully, and the average length of life has been increased spectacularly, but the maximum length of life remains the same. Some cancers can be cured or palliated, but the number of cancer deaths is still[10] 12.5% of all deaths, only exceeded by vascular diseases (39%) and infectious and parasitic diseases (23%). Even so, the media and a few scientists and scientific futurologists get excited about the possibility that death will be banished sometime in the near future[11].

There are, however, many inklings that seem to point to a possible upper limit to the length of human life. Disease itself may never be conquered, as new diseases appear all the time (AIDS is a well-known recent addition). Against this, some scientists assume that the final conquering of death will come when we achieve real artificial intelligence and are able to download our conscience, memories and personal characteristics into a plastic-metallic body and a silicon brain, or equivalent. Again, this question is related to the three previous ones, and therefore may well be outside the scope of science.

Another related problem in this context is the following: would we want to live forever, if it were feasible? The answer to this has been frequently attempted in fictional literature, usually on the negative[12]. We must also take into account that, if immortality were reachable, humans should either stop reproducing or be forced to conquer the galaxy (which would only be a postponement of the problem).

The limits of technology

Again, as in the case of biology, the limits we confront here are practical, rather than theoretical. There are many, but we will focus just on Moore's law, because it is very well known and has to do with hardware and computer science, the most publicized technologies of our time.

Moore's law, which was formulated in 1965 by Gordon Moore[13], can be stated thus:

In 18–24 months the following magnitudes are doubled:

- The number of components in a chip
- The clock frequency
- The memory capacity

and the following magnitudes are halved:

- The price per component
- The energy consumption per component

Observe that the first, fourth and fifth bullets combine to predict that the price of a chip and its total energy consumption remain practically constant with time.

Let us look at an example: in 1965, the price of a transistor was \$1. In 2003, for that price you could buy a chip containing 15 million transistors. In 38 years, the number of transistors in a chip duplicated almost 24 times, which corresponds to a duplication every 1.6 years.

Let us look at another example: Figure 1 shows the evolution with time of the clock frequency used by microprocessors. It can be seen that this magnitude has duplicated over 8 times in 18 years, which corresponds to a duplication every 2.2 years. The rate of growth is clearly exponential.

Similar curves can be drawn for the number of transistors in a microprocessor and other magnitudes. Figure 2, for instance, shows the evolution of DRAM memory capacity. The axis are logarithmic, because a pure exponential growth can be better appreciated in that way (it shows as a straight line).

In 2003, Gordon Moore was invited to address a lecture at an IBM Academy meeting. In that lecture, he made the following statement: *No exponential is forever*, which he later used as the title of a paper in a conference[14]. This is a well-known fact, but it is also something that people tend to

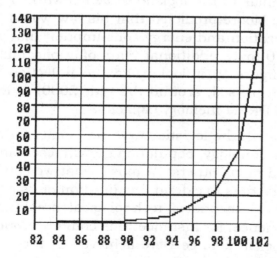

Figure 1. The evolution of microprocessor clock frequency.

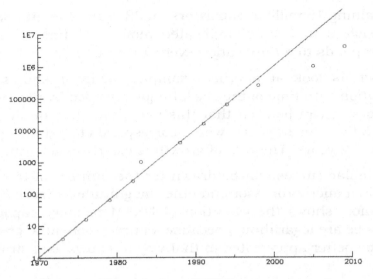

Figure 2. The evolution of DRAM memory capacity.

forget. All natural growth processes start looking like an exponential growth, but sooner or later (when natural limits are reached) the curve starts growing more slowly until it becomes similar to the *logistic curve* shown in Figure 3. In view of this, Moore predicted that his law would find limits in miniaturization and energy consumption that would give it at most 10 to 12 additional years of applicability. In fact, looking at figure 2, it looks like the inflection point has already been passed, around the year 2000, at least in the evolution of DRAM memory capacity.

As indicated, the logistic curve sets limits for all natural processes: technology, population growth, economical development, and many others. At most, what we can expect (as Moore observed) is to extend the duration of its exponential-like section by means of technological breakthroughs. In computer science, for instance, this could be done if a drastic technological change were achieved in the near future, such as quantum or photonic computing, or a possible

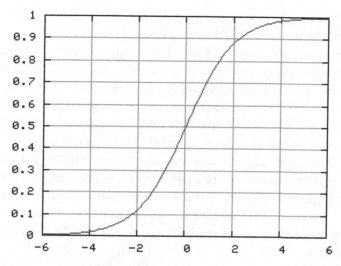

Figure 3. The logistic curve

change from silicon to graphene technology. But this is something we can only surmise at this point and much more work will be needed before it is made feasible.

References

1. To escape from Russell's paradox he replaced Frege's form of set theory by the theory of types, which forbids that a set may be a member of itself.
2. *Über formal unentscheidbare Sätze der "Principia Mathematica" und verwandter Systeme.*
3. *Randomness and Mathematical Proof*, Scientific American 232, No. 5 (May 1975), pp. 47–52, *http://www.cs.auckland.ac. nz/~chaitin/sciamer.html.*
4. The importance of this theorem for providencial evolution was signaled by Fernando Sols, *Heisenberg, Gödel y la cuestión de la finalidad en la ciencia, in Ciencia y Religión en el siglo XXI: recuperar el diálogo*, Emilio Chuvieco and Denis Alexander, eds., Editorial Centro de Estudios Ramón Areces (Madrid, 2012). See also chapter in this book by the same author.

5. The emperor's new mind.
6. *Traité élémentaire de Chimie* (1789).
7. Leptons are a family of particles comprising electrons and positrons, heavy electrons or muons, neutrinos and the tau particle. For leptons, we are still in the third level, since we have not been able to discover any inner structure for them.
8. Bosons are elementary particles with integer spin, which serve as intermediary for fundamental interactions between particles. They include the photon and the W and Z particles, as well as a few undiscovered particles, predicted by the standard model: gluons, the graviton, the Higgs boson, and a few others.
9. In a letter of condolences, Einstein wrote: *The distinction between past, present and future is only an illusion, although persistent.*
10. 2002 figures.
11. 2025 for deaths not caused by accidents, according to Ray Kurzweil in *The singularity is near*, 2005.
12. See *Los inmortales* (*The immortals*), a short story by Jorge Luis Borges included in his collection *El aleph*, 1949.
13. *Cramming more components onto integrated circuits, Electronics* 38(8).
14. G.E. Moore, *No exponential is forever: but "Forever" can be delayed!*, Solid-State Circuits Conference, 2003. Digest of Technical Papers. ISCC. 2003. IEEE International. *http://ieeexplore.ieee.org/xpl/articleDetails.jsp?arnumber=1234194.*

Euclid in Raphael's *School of Athens*

5. On the Intelligibility of Quantum Mechanics

Julio A. Gonzalo

Before quantum mechanics

Quantum Mechanics has the reputation of being not very intelligible. There are, of course, different intellectual approaches to objective reality: e.g. the inductive approach, the deductive approach and the intuitive approach. Physics, as Einstein noted, is "simple but subtle", and the direct, intuitive approach is not always the most fruitful. For instance, the classical concept of "inertial movement" did not occur to Aristotle, one of the most competent and keen observers of nature. It had to wait till Buridan and Oresme, in the 13th century, still a long time before Newton, to be clearly formulated for the first time.

It is true that Quantum Mechanics is often counterintuitive. After first introducing in historical perspective the subject matter of this chapter, we will make some common sense considerations indicative of the fact that we must expect intrinsic limitations in our knowledge of the atomic and subatomic world which is the proper realm of Quantum Mechanics.

In the second half of the 19th century, astrophysics, nothing else but ordinary physics as formulated by Newton and applied to heavenly bodies, was able to predict by means of rigorous calculations the existence of a new planet beyond Saturn, which soon would be called Neptune. Physics had discovered by then the wave properties of light and its electromagnetic nature. It had learned to understand colour, sound, heat and electricity. Physicists and chemists had been able to discover the periodic table of the elements, and had discovered how to do the chemical analysis of light coming from the stars. Physicists had discovered radio waves and X-rays, and they had given the first steps in our understanding of the properties of solids and the phenomenon of natural radioactivity. Einstein would say few years later that *"the most incomprehensible fact about our universe was that it was comprehensible"*.

Nobody in those closing years of the 19th century doubted that classical physics was in possession of the fundamental laws of the universe. A famous theoretical physicist said then that the only remaining task for physicists was to add a few more exact numbers to known results, such as those giving the gravitational constant or the velocity of light in free space.

In those days[1] William Thomson (Lord Kelvin), who had been a prodigious child, a bright student, and then the most respected Physics professor in Great Britain for many years, said that on the whole horizon he saw only "two little clouds": The negative result of the Michelson-Morley experiment, trying to measure the relative motion of the Earth with respect to the "ether", and the ultraviolet catastrophe observable in the "blackbody" radiation spectrum, which did not fit the classical Rayleigh-Jeans formula derived assuming equipartition of energy among all the oscillators contributing to the "blackbody radiation".

This penetrating observation of Lord Kelvin anticipated Planck and Einstein, who would soon call into question the classical concepts of space, time, matter and radiation. In particular the pioneering work of Planck (1900), introducing the concept of *"quantum"* of energy" (hv), followed by its successful application by Einstein (1905) to explain the *photoelectric effect*, was the first step on the road leading to the formulation of Quantum Mechanics, twenty years later, by a phalange of young bright physicists: de Broglie, Heisenberg, Schrödinger, Dirac, Jordan and Born. The rich heritage from Classical Physics (Mechanics, Electromagnetism, Optics, Thermodynamics) was certainly extraordinary, but the successful application of Quantum Physics to understand molecules, atoms, nuclei and elementary particles opened the way to the *technological revolution* of the second half of the 20th century, Solid State materials, devices and sensors made possible many things, from computers to artificial satellites, and permitted us to put a man on the Moon.

Blackbody radiation and Planck's constant

In a single stroke Planck succeeded in accounting for all the main features of blackbody radiation. A blackbody is a cavity filled by radiation in thermal equilibrium at a given temperature. The radiation field occupies the entire cavity and the radiation emitted through small holes on the surface of the cavity has characteristic properties entirely defined by the equilibrium temperature. They are totally independent of the atoms forming the surface of the cavity.

The *main features* of blackbody radiation are[2]:

a) The characteristic *spectral distribution* (*Planck's Law*) given by

$$W_T(\omega)d\omega = \frac{\hbar\omega^3}{\pi^2 c^3} \frac{1}{e^{\hbar\omega/k_B T} - 1} \qquad (1)$$

Where $W_T(\omega)$ is the radiation energy density per unit volume and per unit frequency interval, ω is the frequency, and T the temperature. The *dimensionless ratio*, $\hbar\omega/k_B T$, which takes all possible values from the extremely small (low frequencies and high temperatures) to the extremely large (high frequencies and low temperatures), is modulated by the *prefactor* $\hbar\omega^3/\pi^2 c^3$ involving therefore three *universal physical constants*:

$\hbar$ (Planck's constant) = $h/2\pi$ = 1.05457266 × 10^{-27} ergs·s,

k_B (Boltzmann's constant) = 1.380658 × erg·K^{-1}

c (speed of light in vacuum) = 2.997924 × 10^{10} cm·s^{-1}

b) The *Wien displacement law*, stating that, as T increases, the characteristic maximum in $W_T(\omega)$ defined by Plack's Law occurs at a frequency ω_{max} which displaces itself to higher values with increasing T according to the relation.

$$\hbar\omega_{max} = 2.8 k_B T \tag{2}$$

This law allows immediate characterization of the blackbody emitting body through knowledge of its equilibrium temperature, or vice versa, the value of its equilibrium temperature from the previous determination of the value of ω_{max}. NASA's COBE satellite was able to obtain with extremely high precision the present temperature of the cosmic background radiation as T_0 = 2.726 K.

When COBE'S data were shown to the American Astronomical Society on January 13, 1990, by its principal investigator, John C. Mather, it brought a standing ovation.

c) The *Stefan-Boltzmann total radiation law*, giving the integrated radiation in the whole frequency range of the spectra from zero to infinity,

$$\int_0^\infty W_T(\omega)d\omega = \left[\frac{\pi^2}{15}(\hbar c)^{-3} k_B^4 \right] T^4 = \sigma_{SB} T^4 \tag{3}$$

where σ_{SB} = 7.5659 × 10^{-15} $K^4 cm^{-3}$ is the Stefan-Boltzmann constant, given in terms of $\hbar$ and k_B.

Combining these three equations Planck was able to determine for the first time the value of h, Planck's constant, in 1900.

By the time Planck delivered his Nobel Prize acceptance speech[3], the quantum of energy $\hbar\omega$ had explained much more than blackbody radiation. That much more included the photoelectric effect, the spectrum of hydrogen atoms and the specific heat of solids at low temperatures. Decades later, through the data of the COBE satellite on the cosmic background radiation, Planck's constant was to play a decisive role in making scientific cosmology quantitatively precise for the first time. From the very beginning, Planck envisioned a system of natural units of mass, length and time based upon the universal constants $\hbar$ (Planck), c (velocity of light), and G (Newton), which should be universally valid.

Waves and particles

Albert Einstein (1905) was the first to see the coexistence of wavelike and particlelike features in light quanta, and Louis de Broglie (1923–24) the first to ascertain their coexistence in material particles. After the bold first step of de Broglie, Quantum Mechanics developed very quickly as a rigorous abstract construction eminently successful in describing molecules, atoms, nuclei and elementary particles. The paths of development opened up by Heisenberg and Schroedinger were complementary but not identical. And philosophical interpretation as developed by the Copenhagen School (presided by Niels Bohr) gave Quantum Mechanics a certain halo of impenetrability. Einstein, de Broglie and Schroedinger were never fully satisfied with the Copenhagen interpretation, but due to Bohr's prestige, and to Heisenberg's Concurrence, that interpretation became dominant in academic circles throughout Europe and America. In the prologue to the book "Les quanta" (Ref. 1), de Broglie asks himself: "Could not the localization of the particles be achieved, leading thus to the notion of corpuscle with its unique precise meaning? Would not we be able to achieve this way an analogous progress in Quantum Physics as the progress achieved by Boltzmann and Gibbs one century ago with the abstract "Thermodynamics of Principles", introducing the idea of localized atoms and developing therefrom the "Statistical Thermodynamics" which has allowed the understanding of the true significance of entropy, that leads us to foresee phenomena which had escaped previous theories?".

For the moment, all attempts to reformulate Quantum Mechanics in a more satisfactory, less abstract way, have been unsuccessful, but in Physics one is not entitled to accept a theory as something unquestionably final.

The original approach of the Broglie to the concept of a material particle as a matter wave was that of connecting

$$E = mc^2 \text{ (Einstein)}, \tag{4}$$

where E in the energy, m the mass of the particle, and c the velocity of light, with

$$E = h\upsilon \text{ (Planck)}, \tag{5}$$

where E is energy, h Planck's constant (a universal constant of proportionality with the dimensions of "action" = "energy" x "time") and υ the frequency of the periodic motion associated with the particle.

To satisfy the principle of Relativity, the law connecting relations[4] (4) and (5) must be invariant for all observers moving with rectilinear and uniform velocity. If the mass of the particle has a rest value m_0, m is always $m > m_0$ for any $v > 0$ according to the principle of Relativity. And the observer at rest with the particle must see a characteristic frequency v_0 such that $v > v_0$ for any observer moving with velocity higher than zero.

De Broglie noted that the frequency ν must vary exactly with the same velocity dependence as the mass m and concluded that the particle's wave length would be then given by

$$\lambda = \mathrm{v}T = \frac{\mathrm{v}}{\upsilon} = \frac{\mathrm{v}}{E/h} = \frac{\mathrm{v}h}{m\mathrm{v}^2} = \frac{h}{m\mathrm{v}} = \frac{h}{p} \tag{6}$$

This is the expression for the de Broglie wavelength of a massive particle moving with velocity v in terms of Planck's constant and the particle's momentum $p = mv$. This was the first step forward for Wave Mechanics or Quantum Mechanics.

De Broglie's relationship was soon confirmed by experiments showing the scattering of particles (electrons, neutrons) by crystals, which fulfill

$$2d \sin \theta = n\lambda \text{ (Bragg's Law)} \tag{7}$$

where θ is the angle of scattering (angle between the incident beam and the scattered beam), d is the periodic distance between parallel scattering planes in the crystal, and n is an integer $n = 1, 2,...$, with $n = 1$ given the first order (strongest) scattering intensity.

In Quantum Mechanics material particles are viewed as wave packets in which we can distinguish the phase velocity

$$\mathrm{v}_{\mathrm{phase}} = \lambda \upsilon = \frac{\lambda}{p} \frac{E}{h} = \frac{E}{p} = \frac{1}{2}\mathrm{v}, \tag{8}$$

which is unimportant, and the group frequency,

$$\mathrm{v}_{\mathrm{group}} = \frac{d\omega}{dk} = \frac{d(E/\hbar)}{d(p/\hbar)} = \frac{d(p^2/2m)}{dp} = \frac{p}{m} = \mathrm{v}, \tag{9}$$

which is the most meaningful and gives the particle's velocity.

Heisenberg's uncertainty principle

Bohr's theory of the hydrogen atom, which assumed the quantization of electron orbits at distinct distances from the nucleus, orbits in which the stationary electron would not radiate energy (therefore contradicting classical electrodynamics) was reasonably successful in explaining quantitatively the emission and absortion spectra of the hydrogen atom, but contained evident contradictions. The deficiencies of Bohr's theory were overcome by means of a brand new mechanics, later called Quantum Mechanics or Wave Mechanics which developed through two different approaches: Heisenberg's approach (1925) which followed on Bohr's path, and Schrödinger's approach (1926) which followed more Einstein's and de Broglie's path. We will try to explain the basis of the conceptual structure of Quantum, Mechanics as far as possible without relaying too much on complicated mathematical formulae.

Let us consider a material wave-particle (for instance, an electron) moving in free space. We can associate a position (x) or a time (t) with the material wave-particle, and a momentum (p) or an energy (E) with the same material wave particle. The pairs x, t and p, E change in the same way under a Lorentz transformation according to Special Relativity.

According to Heisenberg's uncertainty principle, because of the intrinsic wave-like character of the electron, when we measure it, there must be an intrinsic uncertainty Δx in the position and an intrinsic uncertainty Δp in the momentum, such that

$$\Delta x \cdot \Delta p \cong h \text{ (Planck's constant)} \qquad (10)$$

For instance, when we scatter a photon over the electron, if it is a very low energy photon it changes the momentum of the electron very little (Δp small), but the photon wavelength is large, and therefore the resulting information produced by the scattering event about the position of the electron is poor (Δx large). On the contrary, if it is a very high energy photon, it may substantially change the electron's momentum (Δp large), but its wavelength is small (Δx small), which is in qualitative agreement with Eq. (10).

Analogous considerations lead Heisenberg to conclude that

$$\Delta t \,.\, \Delta E \cong h \text{ (again, Planck's constant)} \tag{11}$$

Therefore, Heisenberg's principle seems to be a negative physical principle, setting strict limits on the maximum accuracy in the measurement of conjugate pairs of physical magnitudes like (x, p) and (t, E). But this is not at all the whole story. A judicious use of Heisenberg's principle leads to extraordinary unexpected results in the microscopic realm of the atom as well as in the macroscopic realm of the stars. In particular, good introductory university textbooks such as M. Alonso and E. J. Finn, show explicitly how Heisenberg's principle leads to the determination of the ground energy of the Hydrogen atom,

$$E_{\min} \cong \frac{me^2}{2(4\pi\varepsilon_0)^2\hbar^2} \cong 13.6\,\text{eV} \tag{12}$$

and to a quantitative estimate of the number of protons (and electrons) in a typical star like our Sun,

$$N \cong \left(\frac{\hbar c}{Gm^2}\right)^{3/2} \cong 2\times10^{57}, \tag{13}$$

resulting in a mass

$$M \cong Nm_p \cong 3 \times 10^{33}\text{g}$$

Concluding remarks

As it is well known, Classical Mechanics can give exact results for a system involving two massive bodies under the action of the gravitational force, but can deal only approximately with a three body system. It should not be surprising therefore that Quantum Mechanics, which is eminently successful in describing the energy levels of a hydrogen atom (a proton + an electron), needs to recur to all kinds of approximations even for the hydrogen molecule (two protons + two electrons).

Heisenberg's uncertainty principle is sometimes referred to as Heisenberg's indeterminacy principle, implying, arbitrarily, that the intrinsic errors in the measurements of conjugate variables, like $(\Delta x, \Delta p)$ or $(\Delta E, \Delta t)$ are a result of the "shaky" character of reality itself, not of lack of proper means to measure those same conjugate variables.

The evidence for radiation quanta and material particles behaving as wave packets, and the fact that Heisenberg uncertainty principle holds with $\hbar = 1.05 \times 10^{-29}$ erg·s (very small but finite) implies that photons and ordinary particles cannot be consistently considered as pointlike entities or small spheres with a definite center, and[5] cannot be taken as absolutely at rest. If $\hbar$ tends to zero, however, these restrictions would disappear: Δx and Δp could go simultaneously to zero, and ΔE and Δt also. We know that electrons, protons and neutrons in atoms and in nuclei behave in such a way that $\hbar = 1.05 \times 10^{-29}$ crg.s. The energy scale for events involving atoms is of the order of $1eV \cong 1.6 \times 10^{-12}$ erg and that for larger scale reactions is $1MeV \cong 1.6 \times 10^{-6}$ erg.

Big accelerators are able to investigate collisions of elementary particles in the energy range of $1GeV \cong 1.6 \times 10^{-13}$ erg, at speeds very close to the velocity of light. Whether the relevant quantum of action for events in the energy range of 10^{15} GeV or more is the same, much larger or much smaller than $\hbar = 1.05 \times 10^{-27}$ erg·s, we do not know for sure. If it is much smaller the restrictions of our knowledge in Δx and Δp would of course be much smaller, but this would not alter the general picture.

Quantum Mechanics was from the beginning a probabilistic theory. By 1964 the fact that Quantum Mechanics was probabilistic already had a long history. The fact that identically prepared systems can yield different outcomes seems to challenge a basic tenet of science. Frustration with this indeterminacy was shown by Albert Einstein when he said that "God doesn't play dice".

Does this mean that Quantum Mechanics contradicts the principle of causality? No.

As **Stanley L. Jaki** has pointed out[6]:

"Had Heisenberg sensed something about the inherent limitations of the methods of physics, he would have refrained from stating that the inability of the physicist to measure nature exactly showed the inability of nature to act exactly. He would have needed only a modicum of philosophical sensitivity to note that the apparent truth of the foregoing statement depended on taking the same word *exactly* in two different meanings: one operational, the other ontological.

His confusion about those two different meanings meant an indulging in that elementary fallacy which the Greeks of old called *metabasis eis allogenos*. That fallacy was duly pointed out in the courses in introductory logic that were part of the curriculum in philosophy, without which no doctor's degree in any subject could be granted in German universities when Heisenberg received his own in physics in 1925.

In taking Heisenberg's principle for a final disproof of causality, no qualms could be felt by physicists in the Anglo-Saxon world, where empiricism and pragmatism had for some time discredited questions about ontology... While Plack quickly perceived that the Machist interpretation of science threatens confidence in the reality of a causally interconnected universe, he invariably equated ontological causality with the possibility of perfectly accurate measurements.

Again, whereas in his three-decade-long dispute with Max Born, Einstein often came to the defense of reality as such, he never perceived what he really wanted to defend. Nor is it likely that he would have been told about it as bluntly by W. Pauli as the latter put the matter to Born in a letter written on March 31, 1954, in a room or two away from Einstein's own in the Institute of Advanced Study in Princeton.

Einstein's arguments of behalf of a reality existing even when not observed amounted to no more than graphic phrases and gestures. He thought to the end that what he meant to be ontological causality, though he never called it such, could only be saved if perfectly accurate measurements were at least in principle possible. The chief failure of his famous thought experiments with a clock on a spring scale was not that it did not work but that it granted a reduction of the ontologically exact to exact measurability. He could not, of course, expect from Bohr, who refuted that thought-experiment, to be reminded of that reduction, since it was a principle with Bohr to avoid any reference to ontology as such."

References

1. J. Audrade e Silva y G. Lochak, *Los cuantos*, (translated form the French *Les quanta* by M. Alario). (Editorial Guadarrama: Madrid, 1969).

2. See e.g. R. Loudon, *The Quantum Theory of Light*, 2nd ed. (Oxford: Clarendon Press 1983).

3. S. L. Jaki, "Numbers decide or Planck's constant and some constants of philosophy" in Julio A. Gonzalo (Coordinator), "Planck's constant: 1900–2000" (U.A.M. Ediciones: Madrid, 2000) p. 129.

4. See e.g. M. Alonso y E. J. Finn, *Física* (Adison-Wesley Iberoamericana: Argentina. Chile. Colombia. Ecuador. España. Estados Unidos. México. Perú. Puerto Rico. Venezuela, 1992).
5. Y. Aharonov, Sandy Popescu and Jeff Tollaksen, *Physics Today*, November 2010, p. 27.
6. Stanley L. Jaki, *God and the Cosmologists*, pp. 124-26, 130-31. (Real View Books: P.O. Box 1793, Fraser, Michigan, 1998) First published in 1989.

Werner Heisenberg (5 December 1901–1 February 1976)

6. Uncertainty, Incompleteness, Chance, and Design

Fernando Sols[*]

The 20th century has revealed two important limitations of scientific knowledge. On the one hand, the combination of Poincaré's nonlinear dynamics and Heisenberg's uncertainty principle leads to a world picture where physical reality is, in many respects, intrinsically undetermined. On the other hand, Gödel's incompleteness theorems reveal us the existence of mathematical truths that cannot be demonstrated. More recently, Chaitin has proved that, from the incompleteness theorems, it follows that the random character of a given mathematical sequence cannot be proved in general (it is 'undecidable'). I reflect here on the consequences derived from the indeterminacy of the future and the undecidability of randomness, concluding that the question of the presence or absence of finality in nature is fundamentally outside the scope of the scientific method[1].

Introduction

Since the publication in 1859 of Charles Darwin's classic work *On the origin of species*, there has been an important debate about the presence or absence of design in nature. During the 20th century, progress in cosmology has permitted to extend that debate to include the history of the universe. The intellectual discussion has become especially intense in the last few decades after the so-called "intelligent

[*]Reference: Fernando Sols, *Uncertainty, incompleteness, chance, and design*, in *Intelligible Design*, Manuel M. Carreira and Julio A. Gonzalo, eds., World Scientific (Singapore, 2013).

design" has been proposed as a possible scientific program that would aim at proving the existence of finality in biological evolution[2]. In this often unnecessarily bitter controversy, chance plus natural selection on the one hand and intelligent design on the other hand, compete as possible driving mechanisms behind the progress of species. Chance, or randomness, is no doubt an essential concept in many fields of science, including not only evolution biology but also quantum physics and statistical physics. It is surprising however that, within the mentioned controversy, it has been barely noticed that, within mathematics, randomness cannot be proved. More precisely, Gregory Chaitin has proved that the random character of a mathematical sequence is, in general, undecidable, in the sense proposed by the mathematicians Kurt Gödel and Alan Turing[3].

In this chapter we shall argue that Chaitin's work, combined with our current knowledge of quantum physics, leads us inevitably to the conclusion that the debate on the presence or absence of finality in nature is fundamentally outside the scope of the scientific method, although it may have philosophical interest. The argumentation will take us to review some decisive moments in the history of physics, mathematics, and philosophy of science. We will refer to Newton's physics, Poincaré's nonlinear dynamics, Heisenberg's uncertainty principle, the wave function collapse, Gödel's theorems, Turing's halting problem, Chaitin's algorithmic information theory, Monod's philosophy of biology, the intelligent design proposal, and Popper's falsifiability criterion.

The guiding theme of our argumentation will be the attempt to address a fundamental question that is as simple to formulate as difficult to respond: "What or who determines the future?" We hope that the reflections presented here will contribute to clarify the debate on finality, helping to distinguish between established scientific knowledge and the philosophical interpretation of that knowledge.

Practical indeterminacy in classical physics: Newton and Poincaré

In his monumental work *Philosophiae naturalis principia mathematica* (1687), Isaac Newton formulated the law of gravitation and the laws of classical mechanics that bear his name. The study of those laws with the tools of infinitesimal calculus, which he and Leibniz created, led to a deterministic world view in which the future of a dynamical system is completely determined by its initial conditions, more specifically, by the initial values of the positions and linear momenta of the particles involved together with the relevant force laws.

This mechanistic picture of nature took strong roots, favored by the impressive success of Newton's mechanics in accounting for planet motion and ordinary life gravity. Despite not being confirmed by modern physics, determinism still enjoys some adherents.

In the late 19th century, Henry Poincaré addressed the three-body problem and concluded that the evolution of such a dynamical system is in general chaotic, in the sense that small variations in the initial conditions lead to widely different trajectories at long times. The longer the time interval over which one wishes to predict the system's behavior with a given precision, the more accurate the knowledge of the initial conditions must be. Poincaré concluded that the appealing regularity of the two-body problem, exemplified by the motion of planets around the sun, is the exception rather than the rule. The vast majority of dynamical systems are chaotic, which means that the prediction of their long time behavior is, in practice, impossible. Thus we encounter practical indeterminacy in classical physics.

One could still claim that, even if ruled out for practical reasons, determinism could still survive as a fundamental concept. One could argue that the future of nature and the universe, including ourselves, are determined indeed but in such a way that in practice we can make reliable predictions only in the simplest cases. For practical purposes, this determinism could not be distinguished from the indeterminism in which we think we live. Below we argue that quantum mechanics rules out a deterministic world view on not only practical but also fundamental grounds.

Intrinsic indeterminacy in quantum physics: Heisenberg

During the first third of the 20th century, quantum mechanics was discovered and formulated. For the present discussion, we focus on a particular aspect of quantum mechanics: Heisenberg's uncertainty principle. It tells us that, due to its wave nature, a quantum particle cannot have its position and momentum simultaneously well defined. More specifically, if Δx and Δp are the uncertainties in position and linear momentum, respectively, then the following inequality must always hold:

$$\Delta x \, \Delta p \geq h/4\pi, \tag{1}$$

where h is Planck's constant. If we combine Poincaré's nonlinear dynamics with Heisenberg's uncertainty principle, we conclude that, in order to predict an increasingly far future, a point will be reached where the required knowledge of the initial conditions must violate the uncertainty principle. The reason is that the condition $\Delta x \to 0$ and $\Delta p \to 0$ (necessary to predict the far future) is incompatible with the inequality (1). We thus reach the conclusion that, within the world view offered by modern quantum physics, the prediction of the far future is impossible, not only in a practical but also in a fundamental sense: the information on the long term behavior of a chaotic system is nowhere[4]. Noting that non-chaotic systems are rare and always an approximation to reality, we may conclude that the future is open[5].

A remarkable example is Hyperion, an elongated moon of Saturn with an average diameter of 300 km and a mass of 6×10^{18} kg whose rotation is chaotic. Zurek has estimated that quantum mechanics forbids predictions on its rotation for times longer than 20 years[6].

The deterministic view is still defended by the supporters of hidden variable theories. They claim that, below the apparent indeterminacy of quantum systems, there are some 'hidden variables' whose supposedly precise value would entirely determine the future. The experiments by Alain Aspect in the early 80s, as well as the numerous experiments on quantum information performed ever since, all inspired by the seminal work of John Bell, have ruled out the very important class of local hidden variable theories.

Uncertainty vs. Indeterminacy

The wave nature of quantum particles is reflected in the superposition principle, which states that, if A and B are possible states of a system, then a linear superposition of them such as e.g. C = A + B represents another possible state of that system (with some notable exceptions that will not be discussed here). On the other hand, a state A is said to be an eigenstate of an observable (physically measurable quantity) S if this observable has a well-defined value (without uncertainty) in A, for instance s_A. Similarly, let us assume that B is another eigenstate of S with eigenvalue s_B different from s_A. We say that the physical quantity S has no uncertainty in the states A or B taken separately. However, S is uncertain in the state C because it is a linear combination of two eigenstates of S (A and B) with different eigenvalues (s_A and s_B).

In Schrödinger's wave mechanics, the state of a system evolves linearly in a deterministic way. So we can say that the uncertainty about S in the state C evolves deterministically. However, when we perform an ideal measurement of S, i.e., if we "ask" the system (prepared in state C) which value holds for the physical quantity S, then the "answer" (outcome of the measurement) can only cast one eigenvalue of S (s_A or s_B). Quantum mechanics can only predict the statistics of many measurements of the observable S performed on the system always prepared in state C. In a single experiment, we obtain either s_A or s_B. Then the measurement postulate (which must be added to Schrödinger's wave mechanics) tells us that, starting in state C before the measurement, the system is projected onto state A or B depending on whether the result of the measurement is s_A or s_B. This process is referred to as the "collapse of the wave function". So because of the initial uncertainty in the value of S, the final state of the system after one measurement is undetermined.

In summary, the indeterminacy of the future results from the combined effect of the initial uncertainty about an observable in a given state and the collapse of the wave function when a measurement of that observable is made on the system. The evolution of a chaotic system coupled to an environment can be roughly viewed as a succession of quantum measurements, each one with a random outcome.

What or who determines the future?

We have already noted that quantum physics offers a world view in which the future is not entirely determined. This indeterminacy is not practical but fundamental. It permits us to think that our experience of free will may be real instead of merely subjective. If the outcome of a quantum process is undetermined, there is no fundamental reason to deny that possibility in some neural events which may be a macroscopic amplification of microscopic processes where quantum indeterminacy plays an essential role[7].

We proceed with our attempt to answer the question which gives title to this section: "What or who determines the future?" We have ruled out the determinism of the initial conditions. We may invoke some type of design (or finality) that would condition the evolution of a system to the achievement of a previously desired goal.

There is a type of design, which we shall label here as 'internal', that is not particularly controversial. Except for recalcitrant determinists, everybody agrees that there are engineers who design cars and architects who conceive buildings. Even a tidy teenager's room suggests the action of a designer who has ordered it. However, the consensus about internal design does not settle the question because many natural phenomena are not directly induced by a human being.

There is another type of design, which may be called 'external', that is polemic because it suggests the idea of transcendence. If internal design reflects the free action of human beings, external design would reflect what in the theological language would be called the action of divine providence, by which we mean the influence of God on the world without manifestly altering its laws. The picture of an undetermined world leaves room for — but of course does not prove — the existence of freedom and providence. Free will may act through *a priori* undetermined quantum processes which are likely to take place in our brain (see previous section).

The possible physical support for the action of providence is more difficult to delimit, probably because it is more general. However, the ubiquity of chaotic macroscopic systems, ranging from meteorology to the history of the solar system and the fluctuations of the primitive universe, strongly suggests that the long term indeterminacy of those systems is ultimately of a quantum nature and thus intrinsically open to a variety of evolutions. Returning to the example of Hyperion, if we try to predict the detailed rotation of that Saturn satellite in a century from now, we quickly realize that a wealth of drastically different evolutions is possible, all being compatible with the laws of physics and with the most detailed knowledge we may have of its present motion state. As noted before, the reason is that a detailed prediction of Hyperion's rotation in one century from now would require so precise a knowledge of its present state of motion that, even for a system of 10^{18} kg, it would have to violate Heisenberg's uncertainty principle. The information on the long term rotation of Hyperion is nowhere because it has no possible physical support.

Within the context of evolution biology, the concept of 'intelligent design' has been proposed in the last few decades. Its existence would be necessary to explain the emergence of complex biological structures that would be very unlikely *a priori*. Intelligent design is not very different from the external design we have mentioned above. The problem with the intelligent design program is that it is presented as a scientific program despite the fact that, as will be argued later, the questions about finality in nature lie outside the scope of the scientific method. Intelligent design may be an interesting philosophical or theological proposal, but not a scientific one. We will argue that the same can be said about the absence of design.

To explain the apparently unpredictable behavior of complex systems, especially in the context of biological evolution, the concept of *chance* (or *randomness*) is often invoked. Chance can be understood here as indeterminacy without design. Randomness is a ubiquitous concept not only in evolution theory but also in quantum and statistical physics. Quantum mechanics is successful in predicting the statistical outcome of experiments provided that these are performed on identically prepared systems. This requirement cannot be satisfied in biological history or in our personal lives. Quantum mechanics has little predictive power on experiments of uncertain outcome that can be run only once.

Thus, randomness is a useful concept when one studies the statistical behavior of many processes each of which carries some indeterminacy. The problem with randomness is that, as we shall see, it can never be confidently ascribed to any sequence of properly quantified events. The reason is fundamental because it is a consequence of Gödel's theorems, perhaps the most important result in the history of knowledge.

In the coming sections we try to understand the meaning and the epistemological consequences of the impossibility of proving randomness.

Gödel's theorems

In the 1920s, David Hilbert proposed a research program whose goal would be to prove, for (axiomatic) arithmetic and set theory, the following statements: (i) It should be possible to prove the consistency of the axioms, i.e., that they do not lead to contradictions. (ii) It should be possible to prove their completeness, i.e., that all theorems are derivable from the axioms. (iii) It should be possible to prove their decidability, i.e., that there is a general algorithm such that, when applied to any meaningful formula, would stop at some moment with only two possible outcomes, "yes" if the formula can be inferred from the axioms and "no" if it cannot.

The young Austrian mathematician Kurt Gödel tackled the problem and found that Hilbert's expectations where fulfilled for first-order logic. However, when he confronted the case of axiomatic arithmetic and set theory, he found a negative result, to the surprise and deception of many. Specifically, for Peano arithmetic Gödel proved: (i) It is incomplete, i.e., there is at least one formula such that neither the formula itself nor its negation can be derived from the axioms. (ii) The consistency of the theory cannot be derived from the axioms of the theory. (iii) The theory is undecidable, i.e., there is no general algorithm to decide whether a meaningful formula is derivable or not from the axioms of the theory.

These surprising negative results demolished Hilbert's expectations, according to which mathematics would be akin to a mechanical game where, from a finite set of axioms and logical rules, and with sufficient patience and skill (or with the help of a modern computer) all theorems could eventually be proved, and where an algorithm would exist which would tell us whether a meaningful formula is derivable form the axioms. The dream of mathematical certainty and of the systematic exploration of mathematical truths was vanishing forever. Our conception of mathematics becomes not so different from our conception of a physical theory. If a physical theory postulates universal laws with the hope that no experiments will be found that refute it, mathematics postulates a set of axioms with the hope that they will not lead to contradictions.

The English mathematician Alan Turing put Gödel's undecidability on a more tangible ground. Well before computers had been invented, he conceived a mathematical program as a sequence of zeros and ones such that, when fed with an input (itself a sequence of 0 and 1), it would yield an output that would also be expressed as a series of 0 and 1. This ideal concept of universal computer is referred to as Turing's machine. Turing proved that no algorithm exists such that, when applied to an arbitrary self-contained program (which includes the input), always stops and yields one of two possible answers: 1 if the self-contained program stops, and 0 if it does not. That is, Turing proved that the question of whether or not a program goes to a halt is undecidable. In short, 'the halting problem' is undecidable.

Since the work of Turing, other classes of problems have been identified as undecidable. In some cases, the undecidability is suspected but has not been proved. An important example of undecidable problem is that of asserting the random character of a mathematical sequence.

Randomness

Chance is a concept that, in a vague form, has been invoked since remote times. It is used by the ancient Greeks, especially Aristotle, and even in the Bible. Chance can be understood as indeterminacy without design. Randomness is practically a synonym of chance. The term 'chance' has a dynamic connotation and is used more often in the context of history and biology; 'randomness' has a more static meaning and is preferentially used in physics and mathematics. For many practical purposes, the two terms can be taken as synonymous, since the inability to predict the future is directly related to the absence of a clear pattern in the (conveniently quantified) past events. However, they are not always equivalent. For instance, as a result of chance, it is possible, with a low probability, to generate a non-random sequence[8].

The Argentine-American mathematician Gregory Chaitin has given a static but probably more fundamental definition of randomness that applies to mathematical sequences[9], which is not an important limitation if one is aiming at a quantitative description of nature. It is simpler to define the absence of randomness. A mathematical sequence (made, for instance, of zeros and ones) is not random if it can be compressed, i.e., if there is a shorter sequence that, when applied to a Turing machine, yields the longer sequence.

Then one says that the shorter sequence contains the longer sequence in a compressed form. A long sequence is said to be random if it cannot be compressed, i.e. if no shorter sequence exists that determines it.

A canonical example of non-random sequence is, for instance, the first million digits of the number π (up to 10^{13} digits of π are known). Despite its apparent random character, that sequence is not random because a program can be written, of length much shorter than one million digits, which yields number π as the outcome.

Chaitin has proved that the question of whether or not a long number sequence is random, is undecidable, in the sense of Gödel and Turing. No algorithm exists that, when applied to an arbitrary sequence, goes to a halt casting an answer 'yes' or 'no' to the question of whether that sequence is random.

The consequence is that, even if chance or randomness is a useful — even necessary — hypothesis in many contexts, it cannot be ascribed with total certainty to any mathematical sequence and therefore to any physical or biological process. This consideration may not have practical implications, but it has indeed important epistemological consequences. To the extent that chance is understood as indeterminacy without design, it can never be legitimate to present the absence of design as a scientific conclusion. Randomness can be a reasonable working hypothesis, a defensible philosophical proposal, but it cannot be presented as an established scientific fact when questions of principle are being debated such as the presence or absence of design in nature.

Popper's falsifiability criterion

In his work *Logik der Forschung* (1934), the philosopher of science Karl Popper proposed that the demarcation line to distinguish genuinely scientific theories from those which are not, is the possibility of being *falsifiable*, that is, the possibility of conceiving an experiment among whose possible results there is *a priori* at least one that would contradict the prediction of the theory[10]. This means that a theory containing universal statements cannot be verified (as an infinite number of verifying experiments would have to be made) but can be falsified (as a single, properly confirmed experiment is sufficient to refute the theory). Within this picture, all scientific theories are provisional in principle. However, when a theory correctly explains and predicts thousands of experimental facts after decades of collective scientific work, we can practically view it as a correct description of nature. This is the case e.g. of atomic and quantum theories. They started as bold proposals in the early 19th and 20th centuries, respectively, and nowadays we are as certain about them as we are about the spherical shape of the Earth.

The universal statements that compose a scientific theory are proposed from the empirical verification of many particular (or singular) statements, following an inductive process. Those singular statements must express legitimate certainties, in what refers both to the validity of the mathematical language employed and to the claim that they correctly describe reality. If the observation is made that planets follow elliptic trajectories around the sun, one is implicitly assuming that, within a certain level of approximation, one is entitled to associate a set of empirical points with the mathematical concept of ellipse.

It is this latter requirement, namely, that a set of numbers can be associated with a mathematical object, what cannot be satisfied when a supposedly universal statement invokes randomness. As noted above, the reason is that randomness cannot be ascribed with certainty to any mathematical sequence. And here we cannot invoke the qualification 'within a certain level of approximation', since an apparently random sequence like the first million digits of π is in fact radically non-random.

This observation is compatible with the fact that, for many practical purposes, the first million digits of π can be taken as random. However the above described limitation is important when we refer to laws that invoke randomness with a pretension of universality, especially if the link to randomness is used to draw metaphysical conclusions (such as the absence of design in nature), and even more particularly if those philosophical proposals are presented as part of the established scientific knowledge. All this is reconcilable with the fact that randomness is a useful — even essential — hypothesis in many scientific contexts. However, chance is not a scientific datum that can be used to reach philosophical conclusions.

Chance in the interpretation of evolution biology

The scientific evidence in favor of the historical continuity and the genetic relationship among the diverse biological species is overwhelming, comparable to the confidence we have on the atomic theory[11]. However, for reasons we have already anticipated, the same cannot be said of an ingredient that is often included in the description of evolution biology. The problem is not methodological, since, as already noted, chance or randomness is a useful working hypothesis in many fields of science. The problem appears when the concept of chance is considered sufficiently established to take it to the domain of principles, where philosophical ideas are debated.

In his influential work *Le hasard et la necessité (Essai sur la philosophie naturelle de la biologie moderne)* (1970), the French biologist and Nobel laureate Jacques Monod contrasts chance and natural selection as the two driving mechanisms of evolution[12]. Chance is indeterministic; natural selection is deterministic. He identifies chance with indeterminism without a project, but never defines chance in a quantitative form, except for some occasional reference to its possible quantum origin, which does not solve the mathematical problem. The lack of a precise definition does not seem to deter Monod from invoking chance as an essential concept. He presents it as the only possible source of genetic mutations and of all novelty in the biosphere, and then states that chance is the only hypothesis compatible with experience. As I may seem to exaggerate, I reproduce a quote below. After describing some genetic mutations, Monod writes[13]:

"...[mutations] constitute the *only* possible source of modifications in the genetic text, itself the *sole* repository of the organism's hereditary structures, it necessarily follows that chance *alone* is at the source of every innovation, of all creation in the biosphere. Pure chance, absolutely free but blind, at the very root of the stupendous edifice of evolution: this central concept of modern biology is no longer one among other possible or even conceivable hypotheses. It is today the sole conceivable hypothesis, the only one that squares with observed and tested fact. And nothing warrants the supposition — or the hope — that on this score our position is likely ever to be revised."

One may understand that professor Monod was not aware of Chaitin's work on the fundamental non-demonstrability of randomness, which first appeared in the 1960s but seems not to have reached the biologists yet. However, it is more difficult to understand why he apparently ignored two ideas: (i) the old intuition (previous to Chaitin's work) that chance is, if not impossible, at least difficult to prove; and (ii) Popper's falsifiability criterion, knowing that it is difficult to conceive an experiment or observation that yields, as an unequivocal outcome, the absence of chance (at least,

Monod was not proposing one); that is, knowing that the chance hypothesis is non-refutable in practice.

Design and chance lie outside the scope of the scientific method

While discussing the relation between Chaitin's work on randomness and the debate on finality, Hans-Christian Reichel made the following observation[14]:

"Is evolution of life *random* or is it based on some law? The only answer which mathematics is prepared to give hast just been indicated: the hypothesis of *randomness* is *unprovable* in principle, and conversely the *teleological* thesis is *irrefutable* in principle."

This logical conclusion may be viewed as a virtue of design theories, since they cannot be refuted. However, it may also be viewed as a weakness, since, according to Popper's criterion, a theory which is *fundamentally irrefutable* cannot be scientific.

We thus reach the conclusion that, due to the impossibility of verifying randomness in any particular sequence of (conveniently quantified) events, finality cannot be refuted as a general law. This intuitive conclusion is rooted in Gödel's theorems.

Analogously, we may wonder whether design cannot be verified for a particular sequence of events, or equivalently, whether the chance hypothesis is irrefutable. To prove these two equivalent statements, we may seem to lack a fundamental theorem of the type invoked in the previous paragraph. However, the weakness of the chance assumption is not so much that it is *practically irrefutable* when invoked as an ingredient of a general law, but rather that it is *fundamentally unverifiable* when ascribed to any singular event properly characterized by a mathematical sequence. Popper's falsifiability criterion emphasizes that, in order to be considered scientific, a universal statement should be amenable to refutation by the hypothetical observation of a singular event that contradicted the general proposal. However, such a criterion gives for granted that the general law can at least be verified in a finite number of singular cases which provide the seed for induction. This latter requirement cannot be satisfied by the chance hypothesis, for fundamental reasons rooted in Gödel's theorems.

In brief, Chaitin's work on the undecidability of randomness leads us to conclude that the design hypothesis is irrefutable as a general law while the chance hypothesis is unverifiable in any particular case. Both types of assumptions lie therefore outside the reach of the scientific method.

A similar conclusion may be reached following some intuitive reasoning not necessarily rooted in fundamental theorems. To that end, we imagine a debate between two scientists who are also philosophers. Albert is in favor of chance; Beatrice favors design. They are shown two sequences describing two different natural processes. The first one seems random; the second one displays some clear non-random patterns.

They discuss the first sequence, unaware of Chaitin's work on Gödel and Turing. Albert claims that the sequence is obviously random, since it shows no clear pattern. Beatrice responds saying that the sequence is designed, although not manifestly so. She claims that the designer has wished to give the sequence an appearance of randomness. They don't reach an agreement.

Had they been informed about Chaitin's work, the debate would not have been very different. Albert would have continued to claim that the first sequence is random but admitting that he cannot prove it. Alice would have insisted on the ability of the designer to simulate randomness and quite pleased would have noted that the random character of the sequence was unprovable in any case. No agreement would have been reached.

Now they discuss the second sequence. Beatrice claims that, quite obviously, it has been designed, since it shows some clear patterns. Albert counters that the sequence is random, noting that, with a nonzero probability, a randomly generated sequence may happen to show some repetitive patterns. He goes on to point out that, given the physical content of the processes described by the second sequence, those patterns have been necessary for the existence of the two debaters. If the sequence had not shown those regularities, none of them would be there to argue about it. Thus, says Albert, the non-random character of the second sequence should not be surprising, as it is a necessary condition for the very existence of him and Beatrice. Again, they don't reach an agreement.

The two philosophical contenders are unable to reach an agreement and no experiment seems to exist that can settle the question. The debate we have just described is obviously a caricature of a real discussion. However, it is easy to find in it reasoning patterns that are frequently heard in debates about the presence or absence of design in natural processes, whether biological or cosmological. When there is a strong philosophical motivation to maintain an interpretation, there is always an argument to defend it in front of the experimental appearance. It is naturally so, because in the debate about finality no decisive experiment or observation can be conceived.

It is interesting to note that the existence of design is non-controversial in other contexts. Nobody questions the existence of design in an airplane, although strictly speaking it is not more provable or less refutable than in the case of biological evolution. There is no experiment that casts as a result that the airplane has been designed. The difference is that we have an experience of design in ordinary life; we know that there are engineers who design airplanes. However, we don't have similar evidence about the existence of an external designer promoting the evolution of species. Because of this, the question of design in biological evolution will always be more controversial.

We are led to conclude that the debate about the presence or absence of finality lies outside the scope of the scientific method. Returning to our two previous contenders, it is clear that, even if they are able to agree about the apparent random or non-random character of the sequences, the debater in a weak position always has an argument to deny the apparently winning interpretation. We have argued that the apparent irreducibility of the chance-design debate is actually fundamental, since it can be regarded as a consequence of Gödel's theorems.

It seems more constructive that, in their daily scientific work, the two researchers participating in the discussion just described, choose in each context the working hypothesis which best stimulates the progress of knowledge, leaving for the sphere of the philosophical interpretation those considerations about finality which can be debated with the tools of reason but not with the tools of the scientific method.

I would like to thank Gregory Chaitin, Javier Leach, Anthony Leggett, Miguel Angel Martín-Delgado, Javier Sánchez Cañizares, Ignacio Sols, Ivar Zapata and Wojciech Zurek, for discussions on the questions here addressed. Posthumously, I would also like to thank discussions with John Eccles and Rolf Landauer. This acknowledgement implies no agreement or disagreement on their part with the theses here presented. Mistakes and inaccuracies are mine.

References

1. This chapter is based on the longer article by F. Sols, *Heisenberg, Gödel y la cuestión de la finalidad en la ciencia*, in *Ciencia y Religión en el siglo XXI: recuperar el diálogo*, Emilio Chuvieco and Denis Alexander, eds., Editorial Centro de Estudios Ramón Areces (Madrid, 2012).

2. W. A. Dembski, *Intelligent Design: the Bridge between Science and Theology* (1999).

3. G. Chaitin, *Randomness and Mathematical Proof*, Sci. Am. 232, 47 (1975).

4. Rolf Landauer used to say that "information is physical". Without a physical support, there is no information [R. Landauer, *Information is Physical*, Physics Today, May 1991, p. 23]. The latter emerges as the various possible future evolutions become specified.

5. K. Popper, *The Open Universe: An Argument for Indeterminism* (1982).

6. W. H. Zurek, *Decoherence, Chaos, Quantum-Classical Correspondence, and the Algorithmic Arrow of Time.* Physica Scripta T76, 186 (1998).

7. The neurobiologist John C. Eccles (1963 Nobel Prize in Medicine) identified a neural process that might underlie an act of free choice [J. C. Eccles, *Evolution of consciousness*, Proc. Natl. Acad. Sci. USA **89**, 7320 (1992); F. Beck, J. C. Eccles, *Quantum aspects of brain activity and the role of consciousness*, Proc. Natl. Acad. Sci. USA **89**, 11357 (1992)]. Other neuroscientists question the reality of objectively free choice [K. Koch, K. Hepp, *Quantum mechanics in the brain*, Nature 440, 611 (2006); K. Smith, *Taking aim at free will*, Nature 477, 23 (2011)]. In regard to the work by Koch and Hepp, we wish to point out that there is much more in quantum mechanics than just quantum computation.

8. For a detailed discussion, see [A. Eagle, *Randomness is unpredictability.* British Journal for the Philosophy of Science, 56 (4), 749–790 (2005)] and [A. Eagle, "Chance versus Randomness", *The Stanford Encyclopedia of Philosophy (Spring 2012 Edition),*

Edward N. Zalta (ed.), URL = http://plato.stanford.edu/archives/spr2012/entries/chance-randomness/].

9. G. Chaitin, *Meta Math! The Quest for Omega*. Vintage Books (New York, 2005).
10. K. Popper, *The Logic of Scientific Discovery* (1959).
11. F. J. Ayala, *Darwin and Intelligent Design* (2006).
12. J. Monod, *Chance and Necessity: An Essay on the Natural Philosophy of Modern Biology* (1971).
13. The emphases are by Monod.
14. H. C. Reichel (1997). *How can or should the recent developments in mathematics influence the philosophy of mathematics?*, in *Mathematical Undecidability, Quantum Nonlocality and the Question of the Existence of God*, A. Driessen and A. Suarez, eds., Kluwer Academic Publishers (Dordrecht, 1997). Emphases are by Reichel.

Albert Einstein (14 March 1879–18 April 1955)

7. A Finite, Open and Contingent Universe

Julio A. Gonzalo

The universe is *finite* (see e.g., Julio A. Gonzalo, *Inflationary Cosmology Revisited*, World Scientific: Singapore, 2005). It has a finite *mass*

$$M = 1.54 \times 10^{54} \, \text{gr},$$

a finite (but growing) "*age*", since the Big Bang,

$$t_0 = 13.7 \times 10^9 \, \text{yrs},$$

and a finite (but growing) *radius*, since that event,

$$R_0 = 9.96 \times 10^{27} \, \text{cm},$$

The finite mass $M = 1.54 \times 10^{54} \, gr$ is conserved, of course through the cosmic expansion, while the actual time and the actual radius are finite, but increasing and, in an *open* universe, unbounded. We may note that the characteristic radius R_+, for a universe with a total matter mass given by $M = 1.54 \times 10^{54} \, gr$, with $y_+ = \sin h^{-1}(1)$, is given by

$$t(y_+) = \frac{R_+}{|k|^{1/2} \, c} \left\{ \sin h \, y_+ \cos h \, y_+ - y_+ \right\} = 0.365 \times 10^9 \, \text{yrs},$$

$$R(y_+) = R_+ \sin h^2 y_+ = 4.58 \times 10^{26} \, \text{cm}.$$

A *Compton* radius

$$r_c = \frac{\hbar}{mc},$$

and a *Schwarzschild* radius

$$r_s = \frac{Gm}{c^2},$$

120

can be associated with the universe as well as with various finite objects in it, going from finite galaxies to finite elementary particles.

The following Table gives the *Compton* radius and the *Schwarzschild* radius for various massive objects in the universe.

If the Friedmann–Lemaitre solutions of Einstein's cosmological equations correctly describe cosmic evolution, (and they do describe well the thermal history of the universe from the Big Bang to present) the dimensionless product of the Hubble constant times the present "age" of the universe is given by

$$H_0 t_0 = 0.942 \pm 0.065,$$

which is incompatible either with a "flat" universe (k = 0), which requires

$$H_0 t_0 = 2/3,$$

or, more so, with a "closed" universe (k > 0), which requires

$$H_0 t_0 < 2/3.$$

Table 1. Compton and Schwarzschild radii for massive objects

Object	m(g)	r_c(cm)	r_s(cm)	r_s/r_c
Universe	$1.54 \cdot 10^{54}$	$2.26 \cdot 10^{-92}$	$1.14 \cdot 10^{26}$	$1.98 \cdot 10^{-118}$
Galaxy	$\sim 1 \cdot 10^{43}$	$3.5 \cdot 10^{-81}$	$7.41 \cdot 10^{-14}$	$0.47 \cdot 10^{-95}$
Star	$\sim 1 \cdot 10^{32}$	$3.5 \cdot 10^{-70}$	$7.41 \cdot 10^{3}$	$0.47 \cdot 10^{-73}$
Earth	$5.95 \cdot 10^{24}$	$5.85 \cdot 10^{-63}$	$4.43 \cdot 10^{-4}$	$1.32 \cdot 10^{-67}$
Planck m.p.	$2.17 \cdot 10^{-5}$	$1.61 \cdot 10^{-33}$	$1.61 \cdot 10^{-33}$	1
Baryon	$1.67 \cdot 10^{-24}$	$2.09 \cdot 10^{-14}$	$1.23 \cdot 10^{-52}$	$1.23 \cdot 10^{38}$
Electron	$9.10 \cdot 10^{-28}$	$3.84 \cdot 10^{-11}$	$6.74 \cdot 10^{-56}$	$0.56 \cdot 10^{45}$

The universe is therefore "open", finite and unbounded.

Modern physics tells us that we live in an evolving, finite and "open" universe.

Why is the universe what it is according to modern cosmology, and not something else?

No purely physical theory and no purely physical experiment can give a concrete answer.

In other words the universe is "*contingent*", it is not *necessarily* what it is, but it is *really* what it is, and not anything else. The term "*contingent*" implies a *physical* reality which cannot be measured directly, but also a *metaphysical* reality (more intangible but no less real) which can be intellectually recognized.

And a "contingent" universe is a "created" universe.

By the middle of the 19th century, both the Hegelian *left* (Marx and Engels) and the Hegelian *right* (specially the Neo-Kantians) had for a fundamental tenet that the universe (material or not) was *infinite*. Only a few first rate scientists dared to disagree. Among them *Gauss*, the prince of mathematicians, who noted that *Kant*'s "dicta" on categories were sheer triviality, probably having in mind non-Euclidean geometries. The finiteness of matter in endless space was implicit also in the work of *Riemann* and *Zöllner*.

Kant's *claim* that the universe was a bastard product of the metaphysical cravings of the human intellect (put forward to discredit the classic cosmological argument to prove God's existence) was flatly discredited with words and deeds by Planck and Einstein, the two greatest physicists of the 20th century.

Planck, after liberating himself of Mach's tutelage, when he affirmed unambiguously (see e.g. S. L. Jaki, "The road of science and the ways to God" (Chicago U. Press: Chicago,

1978)) his full confidence in the reality of a *causally connected universe.*

Einstein, after finally emancipating himself from Mach's influence, produced the first contradiction-free treatment of the totality of all gravitationally interacting things, which explicitly required a *universe with a finite mass.*

When Planck's son, Erwin, was executed for plotting against Hitler at the end of World War Two, everything seemed to have fallen in ruins around him — home, country, science-. He wrote to a friend (see e.g. S. L. Jaki, Ibidem):

"What helps me is that I consider it a favour of heaven that since childhood a faith is planted deep in my innermost being, a faith in the Almighty and the All-good not to be shattered by anything. Of course his ways are not our ways, but trust in him helps us through the darkest trials"

In 1952, few years before his death, Einstein wrote to his friend Maurice Solovine:

"... I think of the comprehensibility of the world... as a miracle (emphasis added) or an eternal mystery. But surely, a priori, one should expect the world to be chaotic. One might... expect that the world evidenced itself as lawful only so far as we grasp it in an orderly fashion. On the other hand, the kind of order created, for example, by Newton's gravitational theory is of a very different character... Therein lays the "miracle" which becomes more and more evident as our knowledge develops... And here is the weak point of positivists and professional atheists, who feel happy, because they think that they have preempted not only the world of the divine but also of the miraculous..."

As Stanley L. Jaki points out in his *The ways to God and the roads of science,* Planck and Einstein, with their confidence in the reality of a causally connected and finite universe, provide extraordinary compelling evidence in favour of a realistic metaphysics and epistemology, midway between idealism and positivism.

Letter to Physics Today[3]

Dear Editor:

As usual, the "Search and Discovery" piece by B. Schwarzschild (*Physics Today, Dec. 2011, pp. 14–17*) on the 2011 Nobel prize was compact and informative. In Fig. 2b a picture with confidence contours for $\Lambda > 0$ in the Ω_m ,Ω_Λ plane shows convergence near $\Omega_m = 1/4$, $\Omega_\Lambda = 3/4$, using information inferred from high-z type supernovae, cosmic microwave background and baryon acoustic oscillations. It implies that there may be three times as much dark energy density as visible plus dark matter energy density.

I would like to point out that an alternative interpretation is possible using the Ω_m, Ω_k plane (where $k < 0$ stands for space-time curvature) instead of the Ω_m ,Ω_Λ plane. It is well known that Friedmann's solutions of Einstein cosmological equations for closed ($k > 0$), flat ($k = 0$) and open ($k < 0$) universes were obtained assuming $\Lambda = 0$ (probably because Friedmann realized that $k < 0$ and $\Lambda > 0$ play similar roles). Hence $\Omega_m + \Omega_k = 1$ in the Ω_m, Ω_k plane results, in exactly the same downwards straight line from (0,1) to (1,0) as it does (in Fig. 2b) in a flat ($k = 0$) universe. It is easy to check that using the Friedmann–Lemaître solutions for an open universe with $\Lambda = 0$, both Ω_m and Ω_k are *time-dependent*. Therefore the straight line from (0,1) to (1,0) describes cosmic evolution from $\Omega_m \cong 1$ (the big bang) to $\Omega_m \cong 0.044$ (now, $t = 13.7$ Gyrs) and beyond, with $\Omega_m = 0$ in the distant future.

Light coming from high-z supernovae with $z \cong 1$ is reaching us *now*, but it may have been emitted at a time

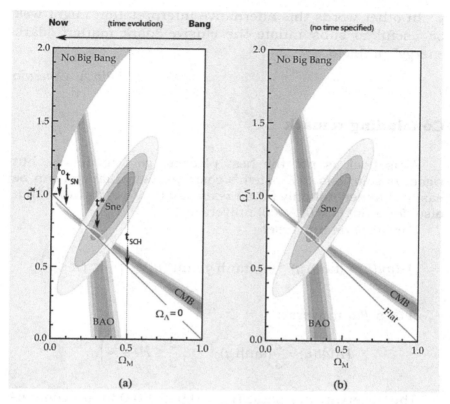

Figure 1. (a) Ω_k vs. Ω_m, and (b) Ω_Λ vs Ω_m

somewhere before $t = t_{SN} \cong 7.5$ Gyrs ago, corresponding to $\Omega_m(t_{SN}) \cong 0.10$, and so forth for more distant supernovae. Beyond some $z < z_+ = 20.7$ (cosmic Schwarzschild radius) galaxies (no stars) are observed for obvious reasons. To correlate quantitatively $\Omega_m(t)$ with $z(t)$ and with t itself is presently straightforward, taking into account that the relevant cosmic quantities ($T_0 = 2.726$ K $\pm 0.01\%$, $t_0 = 13.7$ Gyrs $\pm 2\%$, $H_0 = 67$ Km s^{-1} Mpc^{-1} $\pm 5\%$) are known with fair precision, after the COBE and WMAP satellites, as well as knowing of the present cosmic equation of state (RT = const.) and that radiation and matter densities become equal at atom formation ($t_{af} \cong 1$ Myrs, $T_{af} \cong 3000$ K, $H_{af} \cong 6.48 \times 10^5$ Km s^{-1} Mpc^{-1}).

In other words this alternative interpretation might well be useful to substantiate the elusive "dark matter"–"dark energy" problem.

Julio A. Gonzalo

Concluding remark

This book is not the best place to go into details, but rigorous solutions of Einstein's cosmological equations can be easily obtained not only for an *open* (k < 0, Λ = 0) universe, but also[4] for a *flat* (k = 0, Λ > 0) universe.

For an *open* universe:

$$H(y)t(y) = (\tanh y)^{-2} - y(\tanh y \sinh^2 y)^{-1} \quad \left(\frac{2}{3} \leq Ht \leq 1\right)$$

For a *flat* universe:

$$H(y)t(y) = \frac{2}{3}(\tanh y)^{-1} \quad \left(\frac{2}{3} \leq Ht \leq \infty\right)$$

The observational data, $H_0 t_0 = 0.942 \pm 0.038$ are compatible with both expressions, but cosmic thermal history seems to be much more in keeping with the time evolution described by an *open* universe.

References

1. Sec. Julio A. Gonzalo *Cosmic Paradoxes* (World Scientific: Singapore, 2012).
2. Julio A. Gonzalo *Inflationary Cosmology Revisited* (World Scientific: Singapore, 2005).
3. The letter was not found of enough interest for publication by the Editors.
4. Julio A. Gonzalo (to be published).

This section has been adapted from previously published book by the author in Julio A. Gonzalo and Manuel M. Carreira, *Everything Coming Out of Nothing* (Bentham Science, 2012).

Part II. On the Origin and Development of Life

8. A Brief History of Evolutionary Thought[1]

Thomas B. Fowler and Daniel Kuebler

Speculation and theorizing about organic evolution have a long history, stretching back to Ancient Greece. Six major phases can be identified:

I. Up to 1650   *Ancient speculation*: guesses and hunches regarding possible evolutionary scenarios; domination of special creation and young earth ideas.

II. 1650–1800   *Emergence of modern science*: recognition of fossils as extinct life forms and formulation of first scientific theories about geology and the history of life; continued domination of special creation idea.

III. 1800–1859   *Laying the foundations*: development and refinement of ideas about changing life forms; concomitant development of geology and other sciences; recognition of Natural Selection as a process.

IV. 1859–1910   *Darwin's triumphal entry*: publication of the *Origin of Species*, spread of Darwin's ideas, early battles over evolution and growing doubts.

V. 1910–1960   *"New Synthesis" period*: Incorporation of genetics into evolution and restructuring of evolution around mathematical population genetics; domination of Neo-Darwinism.

VI. 1960–present   *Modern battles over evolution:* rise of modern Creationist movement; critiques of key evolution ideas by Intelligent Design, Meta-Darwinian schools; disagreements within Neo-Darwinian school.

I. Ancient speculation (to 1650)

Ancient Greeks

Two schools of thought about nature appeared in Ancient Greece: believers in the relative immutability of species, and those who advocated a more flexible view. In the first camp was Aristotle (384–322 BC) who systematically attempted to understand nature through first-hand observation. Though not an experimenter, his meticulous observations included a multitude of dissections and led to a very forward-looking scheme for classifying organisms. His anatomical studies led him to maintain a very holistic view of organisms,[2] which naturally led to his belief in the relative immutability of species.

Early Evolutionary Speculation

The opinions of two other ancient philosophers, Anaximander (c. 610–546 BC) and Lucretius (99–55 BC), stand in contrast with Aristotle's view. Both believed that species were quite mutable and, for the most part, were randomly assembled. Anaximander believed that life initially began in the sea with simpler forms giving way to more complex forms. He proposed that humans arose inside of fishes, were reared, and moved onto dry land. His ideas were rather fanciful and no mechanism for evolution was ever discussed, but many give him credit for being the first full-fledged evolutionist.

Lucretius, the first century BC Roman philosopher and poet, can be credited with the first well-developed evolutionary story. In his famous poetic work *De Rerum Natura* (On the Nature of Things), Lucretius struck upon the idea of survival of the fittest or natural selection as the driving force behind evolution. Though he had no idea how this would actually work, his reasoning is remarkably similar to Darwin's ideas.[3]

II. The Emergence of Modern Science (1650–1800)

Early Fossil Discoverers: Extinct Species and the Implied Change in Nature

Despite the antiquity of ideas surrounding evolution, hard evidence took much longer to surface, especially since science and popular culture were dominated by a Biblically informed worldview, centered on special creation. It was not until the 17th and 18th centuries, when scientific efforts were redoubled across Europe, that problems with special creation emerged. The two disciplines with the greatest impact on the advancement of evolutionary ideas were paleontology and geology

Although fossil finds were abundant prior to and during this time, many believed that fossils actually had grown within rocks; otherwise marine fossils would not be found in dry mountainous regions. Leonardo da Vinci (1452–1519) was one of the most vocal opponents of this view, believing it ludicrous to assume that fossils could grow trapped inside rocks without food or air. However, it was not until the mid 1600s that the brilliant Danish biologist/geologist Nicolaus Steno (1638–1686) demonstrated conclusively that specific fossils were the remnants of once living organisms.[4] But unrecognizable fossil forms were still a source of confusion and debate.

Debate over fossils continued for centuries because of two fundamental questions fraught with theological implications: (1) how did the fossils come to be buried within the rock layers? And (2) what exactly did the unrecognized fossil forms represent? Many researchers, Steno included, believed that various fossil layers were formed in successive floods, the most notable of which was the Biblical flood of Noah (Genesis 7) — a theory still advocated by Creationists. This belief was reinforced by the presence of fossilized fish and mussels on arid mountaintops. Unfortunately, the question of the age of the rock and fossil layers was a problem that science at the time was not equipped to address.

The presence of unrecognized forms was approached in two different ways: (a) Dismiss them as aberrations, spurious chance mutants or templates used by God in the creation of real species. (b) Take them at face value, but postulate that they represent species still present in remote unexplored regions of the earth. This was the approach of English naturalist John Ray (1627–1705), considered the father of modern systematic zoology. Although Ray could not accept the position that species had gone extinct, his contemporary Robert Hooke (1635–1703) did so and went one step further, postulating that species had not only gone extinct but that new species had been created.[5]

Hooke's ideas found validation in the work of the father of vertebrate paleontology, the renowned French comparative anatomist Georges Cuvier (1769–1832). To Cuvier and many of his contemporaries, the notion that species had gone extinct was an empirical fact. Cuvier believed that organisms were interdependent wholes resistant to any amount of substantial change, as expressed in his principle of the *correlation of parts*.[6] In such delicately balanced systems, alterations would compromise the functionality of the entire system, compelling species to "keep within certain limits fixed by nature;"

therefore evolution could never occur — an argument still employed by critics of evolution today. Cuvier also argued against evolution, particularly gradual evolution, based on the fossil record, namely lack of transitional forms — another argument still used. He believed that the discontinuous fossil record better corresponded to catastrophic extinctions followed by God repopulating the world with new, distinct species.

Species Change

The father of taxonomy, Swedish naturalist Carl Linneaus (1707–1778), first published his famous classification of living things, the *Systema Naturae*, in 1735. Linneaus was responsible for the convention of naming an organism by its genus and species, and for establishing a comprehensive hierarchical classification scheme which grouped organisms into species, genera, orders, classes and kingdoms. However, the purpose of his classification scheme was not to uncover evolutionary relationships. But using his results others interpreted the fossil record, the success of hybridization experiments, and the lack of clear morphological delineations between certain species and sub-species as evidence that evolution had occurred. For example, based upon this type of evidence, the French naturalist Georges-Louis Leclerc, Comte de Buffon (1707–1788), maintained that species had been altered significantly over long expanses of time by environment changes.[7]

Functionalism vs. Formalism

Cuvier's one time friend Etienne Geoffroy St. Hilaire (1772–1844) focused on the overall form of organs instead of the particular interactions among parts. Rather than seeing the human arm as merely a complex system of interacting parts (functionalism), he saw it as similar in form to arms of other primates, legs of horses, wings of bats and fins of fishes (formalism). What Geoffroy saw was a single archetypal form, which could have been modified to flourish in various environments.[8] Cuvier disagreed, maintaining that structures were only similar in form to the degree that they had a similar function. Thus to Cuvier, the forelimbs of various vertebrates only implied a general similarity of function rather than an evolutionary relationship as Geoffrey argued.

Lamarckian Inheritable Change

Another major influence on Darwin's thinking[9] was the French botanist Jean-Baptiste Lamarck (1744–1829), who formulated the first plausible and well-developed (albeit incorrect) theory of evolution. Lamarck's theory has many similarities with Darwin's, though the mechanism of change is quite different. Lamarck believed that as organisms struggle to adapt to changing environments, they use certain structures or parts preferentially. Lamarck believed these parts would be altered for the better during an organism's lifespan based upon the innate efforts of the organism. He further believed that these alterations could be passed on to the organism's offspring. This is termed *inheritance of acquired characteristics*. Over several generations, it would lead to marked improvements.[10] Lamarck's theory was based upon an idea that was to become a central tenant in Darwinian theory, namely that organisms could evolve as a result of the accumulation of *gradual beneficial* changes. Lamark's theory was discredited when the mechanism of inheritance was rediscovered at the beginning of the 20th century.

Paley and Design

The school of "Natural Theology" held that an understanding of nature inevitably led to belief in and understanding of the Creator. While this school has its roots in the writings of 17th century naturalist John Ray, it was most powerfully articulated in the works of William Paley (1743–1805). Paley advocated that species were distinct entities designed and specially created by God, and that God had taken care to give organisms the adaptations needed to prosper in their environment. Paley believed that if an all-powerful God had taken care to establish the proper arrangement of nature, a blind evolutionary process could only result in disaster. Paley's ideas made a great impression on the young Darwin.

III. Laying the Foundations (1804–1859)

Lyell, Malthus and Blythe

Typical of the much more integrative approach to knowledge in the 19th century (as compared to today) is the fact that two of the ideas with particular impact on Darwin's theory came from outside the realm of biology. The first was from the British geologist Charles Lyell (1797–1875), the principle of *uniformitarianism*: geological features are predominately the result of slow forces of the same kind and intensity as those found acting in the present day, rather than catastrophic floods. Because uniformitarianism demanded long time spans for creation of today's impressive geologic features, Lyell (and subsequently Darwin) advocated a much older earth — tens to hundreds of millions of years older than the proposed Biblical age of the earth. As a geologist, Lyell realized that species had gone extinct and that new species had taken their place, but he regarded the absence of intermediary forms as problematic to both Lamarck's and ultimately to Darwin's ideas. Echoing earlier thought, he believed that gradual changes could improve existing biological structures, but that slow accumulation of slight variations could not account for novel structures.

The second major non-biological influence upon Darwin's thinking came from the work of the British political economist

Thomas Malthus (1766–1834). Malthus observed that many animals and plants produce more offspring than can possibly survive,[11] which leads to fierce competition among the off-spring because the dwindling food supply cannot support them all. When this happened in the human population, Mathlus believed it led to the poverty and poor nutrition, and his solution was forthrightly eugenic: reduce childbirth among the lower classes. However, it was Malthus' observations, rather than his "solutions," that garnered Darwin's attention and admiration, and which he subsequently conjoined with the notion of favorable variations to create his theory.

In addition to these influences from outside biology, it is probable that Darwin was influenced by the British naturalist Edward Blyth (1810–1873) who gave the notion of natural selection a very clear statement in the 1835 edition of the *Magazine of Natural History*, a periodical known to have been read by Darwin.[12] Despite his formulation of how natural selection might work, Blyth believed natural selection to be a profoundly conservative principle, one which seeks to optimize the "form" of a species with respect to its "natural habits", i.e., its environment. Evolution in the stronger sense was considered but rejected by Blythe, Lyell, and others partly because of the fossil evidence, but also because of perceived limitations of species to adapt to their environment near its boundaries. Surely that would be less difficult than the great changes needed to explain the origin of higher taxa such as families, orders, classes, and phyla.[13] For Blythe, this fact about species demonstrated that their capacity to adapt was far too limited to account for the panorama of natural history.

IV. Darwin's triumphal entry and early battles over evolution (1859–1910)

The scientific landscape in the mid 19th century was thus primed for Darwin's simple yet controversial theory.

Key ideas including competition for survival, natural selection, gradual imperceptible change accounting for the complexities of the natural world, and the adaptive fit organisms have for their environment, are found in the works of his immediate predecessors. Darwin synthesized these lines of thought, relying on the large amounts of data he gleaned collecting and examining fossils and specimens throughout the globe. Two things in particular struck Darwin during his travels: (1) Living creatures in a specific geographical area are often closely allied to fossil specimens from the same region. This he took as strong evidence for descent with modification. (2) Radiation of similar types is observable in the different regions of South America. Darwin began to formulate his theory of evolution in his notebooks during the late 1830's but remained hesitant to publish his ideas.

By 1855 Alfred Russell Wallace (1823–1913) had independently struck upon a similar theory (although not as well developed) and was ready to publish his work. In 1858, at the urging of Lyell, a sketch of Darwin's work and Wallace's paper were presented jointly to the Linnaean Society of London. Despite the joint presentation, it was Darwin who would become inextricably linked to the theory of evolution due to the publication one year later of *On the Origin of Species by Means of Natural Selection, or the Preservation of Favoured Races in the Struggle for Life*, a manuscript that laid out his theory in much more detail and forever changed science.

Darwin's contribution was twofold. First, he thrust a new function onto natural selection: the promoter of innovation, so that working in conjunction with random changes in hereditary material, natural selection would choose those changes that were beneficial to the organism.[14] Second, Darwin drew upon his enormous storehouse of first-hand knowledge about organisms and their environments for insight and examples.

For all the impact that it has had on science, it is remarkable that the *Origin* contained no experiments. Rather, it comprised a simple and elegant thesis, supported by a mountain of circumstantial evidence and observations. The thesis Darwin proposed in the *Origin* was as follows (1) organisms within a species vary and this variation is heritable; (2) organisms struggle within their environment for survival; (3) those with advantageous variations will survive and reproduce and will be disproportionately be represented in the next generation. Darwin inferred (or rather extrapolated) that those propagated variations eventually accumulate and result in new species and higher taxa.

Although Darwin admitted that only limited change had been observed in animal husbandry, he posited that this process could be extrapolated over vast amounts of time, so that monumental change could occur.[15] Of course, the mere fact that breeders observe limited amounts of change in species during a few generations, in no way *proves* that large amounts change can occur over long periods of time (macroevolution). Not surprisingly, this was an early objection to the theory, and remains so to this day.

Despite the impressive amount of circumstantial evidence Darwin compiled in support of the theory, it still had problems. With characteristic honesty, Darwin acknowledged these problems, such as the lack of transitional forms in the fossil record, the prolonged time needed for such a process to take place, the difficulties in assembling novel structures through a gradual process, and the sudden appearance of fossil forms. He even admitted that the absence of fossil forms before the sudden explosion of the Cambrian period was a major problem.[16] On the positive side, Darwin, a good scientist, was bold enough to make predictions based on his theory, predictions that he felt future research would confirm. Some of these predictions were later verified; others were not.

Darwin's Critics

The events following the publication of the *Origin* were rather tempestuous. Darwin was attacked, both inside and outside of science. Many of Darwin's scientific contemporaries, although sympathetic to his naturalistic bent, had serious concerns regarding the validity of his theory on empirical grounds. The most significant problem raised was the lack of gradual transitions within the fossil record and within extant species. Darwin argued that the discontinuous nature of the fossil record was an artifact of its spotty nature. He explained the lack of living transitional forms in nature forms by saying that intermediate forms would eventually go extinct as they competed against their more robust cousins. Over time they would disappear all together leaving large gaps between various living forms.

Although these explanations were reasonable, they were highly speculative and many critics were unpersuaded. Darwin's "Bulldog", Thomas Huxley (1825–1895), remained convinced that there was a dearth of evidence for any real transition forms between natural groups. He was of the opinion that evolution was driven by leaps or saltations, an opinion he famously made known to Darwin in a letter.[17]

Another significant problem raised was Darwin's generalization from artificial selection to natural selection. The traits that breeders varied were things like leg length, body size, and coat color — hardly the characteristics that define new families and genera. Traits such as hoofs turning to claws or a heart losing or gaining a chamber were never observed. Thus if artificial selection could only alter *superficial* traits, there was no reason to assume that natural selection should be able to alter *complex* ones. Darwin's theory also ran into problems regarding the supposed young age of the earth, far too young for evolutionary mechanisms to work, though this was soon resolved with the further development of astronomy and physics.

Darwin and Heredity

Darwin's theory tenuously weathered all criticism during his lifetime, but as the 19th century drew to a close, new challenges began to mount, particular those dealing with the mechanism of inheritance. For Darwin's theory, the predominate genetic theory of the day, blending inheritance, was problematic because if blending occurred to a significant extent, small advantageous changes in one individual would gradually be lost though blending with individuals that lack that change. Instead of the gradual spread of minute beneficial mutations, there would be a rapid suppression and elimination of all slight changes, making Darwin's theory highly improbable. This was, in fact, pointed out to Darwin in the 1860s.[18]

Because of these difficulties, Darwin's theory needed a unit of inheritance that was 1) permanent, 2) independent of the unit inherited by the other parent, and 3) inherited in a regular pattern. Rediscovery of Augustinian monk Gregor Mendel's (1822–1884) work on genetics at the dawn of the 20th century was able to provide Darwin's theory with just such a unit. However, in doing so, it raised a whole new set of issues.

Continuity Or Discontinuity?

Mendel found that physically discrete units, which were later termed *genes*, governed inheritance. These genes did not blend into oblivion, but in fact propagated unaltered from generation to generation (except for rare cases of mutation) independently of the corresponding gene from the other parent. While Mendel's work provided Darwin's theory with some much-needed empirical support with regard to heredity, it also demonstrated the discontinuous nature of some heritable traits. It was this revelation that proved to be a stumbling block for the theory. Mendel's data seemed to argue *against* Darwin's idea of gradual continuous variation and supported the notion that evolution had occurred through the leaps and bounds associated with discontinuous variation or saltations.

Though overlooked at the time, Mendel's work was rediscovered by the Dutchman Hugo de Vries (1848–1935), among others, and it was he who most clearly realized the importance of Mendel's work for the study of evolution. De Vries believed that gradual change was no basis for developing a theory of evolution. Rather, speciation in nature must arise by a different process, saltational change, which he believed to be random, plentiful and able to proceed in any direction. Natural selection was relegated to the secondary role of merely sorting through the saltations that had occurred and discarding the disadvantageous ones. The Englishman William Bateson (1861–1926) came to favor a similar saltational theory as a result of his studies as a naturalist.

But despite such criticisms, by the turn of the 20th century naturalistic evolution was triumphant, and all theories of special creation had been soundly routed, at least among the academic community. The only "real" question about evolution then remaining for academics was whether Darwinian gradualism or some new saltational theory would emerge as the comprehensive explanation for the history of life.

V. Genetics and The "New Synthesis" Period: 1910–1960

The early 1900s probably marked the low point in the history of Darwin's theory. For a brief period it appeared that Darwin's notion of gradual evolution would be discarded and replaced by the type of saltational theory favored by de Vries and Bateson. By the 1950's though, Darwin's ideas had been skillfully integrated with a modern understanding of

Mendelian genetics, which enabled credible responses to be given to many objections, and opened new areas for exploration. This monumental achievement was the work of a number of dedicated and gifted men in the first half of the 20th century, and their efforts became known as the "New Synthesis", or what is commonly referred to today as *Neo-Darwinism*. The work of the great American geneticist Thomas Hunt Morgan (1866–1945) in genetics turned that field from a stumbling block into the discipline most responsible for establishing Neo-Darwinism as the reigning orthodoxy.

Morgan began his scientific career skeptical of the Darwinian position. But his research on fruit fly inheritance convinced him that evolution proceeded in a gradual manner working on minor mutations. He came to believe that the gradual accumulation of mutations could improve or modify organisms to the extent needed to drive evolution, and that this process could be described by Mendelian genetics. Morgan never actually witnessed the transformation of a fruit fly into a new species via the accumulation of minor mutations, only the emergence of new variants of a particular species of fruit fly. Even so, based upon these observations, he extrapolated that if minor mutations were allowed to accumulate indefinitely, new species would arise.

In the coming decades every other discipline that studied evolution — paleontology, anatomy, and embryology included — had to come to terms with the burgeoning field of genetics. What really mattered in the study of evolution were the genetic changes that occurred at the chromosome level, and inferring this via paleontology or gross anatomy was impossible.[19] This marked a critical turning point in evolutionary theory and biology in general. The study of organisms via embryology, zoology and botany was forced to take a back seat as genetics and the study of genes began to dominate the evolutionary debate.

Population Genetics

While Morgan established the heritability and spread of relatively minor genetic mutations in the lab, there was no evidence that the accumulation of such mutations in the wild could lead to the origin of new species, much less high taxa. Since this is the fundamental question of evolutionary biology, a vigorous examination of the ability of accumulated minor mutations to bring it about was needed. Rather than focusing on this question — obviously a difficult problem — researchers instead focused their efforts on an easier problem, namely how minor mutations, such as those Morgan observed in laboratory populations, could spread through a population in the wild without human intervention. To answer this question, an entirely new discipline at the border of mathematics and biology emerged known as *population genetics.*

The earliest of the population geneticists was Sir Ronald A. Fisher (1890–1962). Using mathematical models to support his ideas, Fisher argued that all evolutionary changes were either adaptive and maintained in the population, or were maladaptive and eliminated. In addition, Fisher argued that nearly all evolutionary change was of a small magnitude. Fisher believed this because his mathematical models demonstrated that large mutations would have a very low probability of improving the organism, mainly because the large change would be too drastic for the organism to handle. Small changes were much more likely to impart improvement as they would be easier to assimilate into the organism without a massive disruption of its form and function. Fisher did not address the question of whether there are limits to what incremental change can accomplish, as many breeding experiments suggested. Fisher simply assumed that there was always room for improvement by means of small change (as opposed to saltations).

Fisher's contemporary J. B. S. Haldane (1892–1964) also believed that the spread of minor adaptive mutations through a population could lead to new species, but he did not reckon this sufficient to explain every aspect of natural history. In populations many features arose that appeared to be non-adaptive, i.e. they gave the bearer no clear advantage. Such features, which often were used to distinguish between two species, could not have arisen as an adaptive mutation that spread through the population as Fisher proposed.[20] But rather than abandon Darwinism, Haldane allowed room for other factors to influence evolution, such as internal developmental constraints, the limited amount of variation found in nature, and possible rapid speciation via hybridization. In this manner, Haldane was not compelled to find an adaptive function for every characteristic of an organism.

Haldane also differed from Fisher in his view on the *tempo* of evolution, favoring a rapid discontinuous pace over a slow and continuous one.[21] Although both Haldane and Fisher's models are mathematically viable, determining which (if either) accurately describes what happens in nature has proved difficult. Despite more widespread acceptance of Fisher's views, the exact tempo of evolutionary change is still a matter of considerable debate.

The third of the founding fathers of population genetics and by far the most controversial was Sewall Wright (1889–1988). Wright's *shifting balance theory* claimed that evolution occurred in sub-populations with different gene sets competing against each other. Such subpopulations could form randomly, and this random drift would allow novel combinations to be tested by natural selection working on them.[22] In order for such a small group to survive in a changing environment, Wright believed it imperative to periodically reestablish gene flow with the larger population.[23] The notion that groups rather than genes or organisms were the subjects of selective pressure was antithetical to Fisher, and also served to turn Wright and Fisher into lifelong rivals. Fisher's view was the one that eventually won the day and became part and parcel of modern Neo-Darwinian theory, so much so that any whiff of group selection was seen as heretical. Recently Stephen J. Gould (1941–2002), David Sloan Wilson and others have attempted, with some success, to bring higher levels of selection, such as group and species selection, back into mainstream evolutionary thought.

Fisher, Haldane and Wright all published their ideas in the 1930's, and the mathematical precision of their genetic models set the stage for the synthesis of modern Darwinian thought during the 1940s. All three assumed that accumulation of small changes could lead to the formation of new species although their mathematics has been challenged recently by Hoyle and others.[24] The models they proposed demonstrated how genes moved through populations based on their selective advantage. While this was and remains extremely important work, the "genes" they investigated were not actual genes, but rather mathematical abstractions. Hence their models do not take into account many important aspects of real organisms, such as how new genes affect development, morphology, or anatomy — all extremely critical parameters in terms of survival. Because of this inherent limitation, their work furnishes no *biological* proof that small mutational change, mediated by natural selection, can account for novel structures, much less new species or higher taxa. For that, other real-world studies are needed.

The "New Synthesis"

In the late 1930's, there was a call to synthesize the approaches of all the disciplines investigating evolution. In 1942 the Committee on Common Problems of Genetics, Paleontology and Systematics was formed to allow dialogue between the various disciplines; and although the committee finished its work seven years later, the development of the modern evolutionary thought, or what came to be known as the "modern" or "New Synthesis," continued rather informally through the 1960s.

The principal items incorporated into the New Synthesis were the following:

- Populations (population genetics) as the foundation for species change
- Random mutation induced by radiation, chemicals, and various chance factors as the basis of new or changed genetic information
- No inheritance of acquired characteristics
- No saltations

What emerged from the synthesis was rather restrictive. Instead of an expansive theory integrating ideas from many disciplines and open to new interpretations of the evidence, the "New Synthesis" developed a narrow, hard-line view of evolution (Neo-Darwinism), dominated by the work of population genetics.[25] Many observers from other disciplines (paleontology and systematics) complained that the mathematical theory was being used to inform the biology, rather than the other way around.

The ideas and the hardening of the modern synthesis can be reflected in the careers of two of its towering figures, the geneticist Theodosius Dobzhansky (1900–1975) and the naturalist Ernst Mayr (1904–2005). Dobzhansky, a student of Morgan, initially advocated large-scale chromosomal rearrangements and random drift as the predominant factors in evolution. Over time he began to discount these mechanisms and gave primacy to natural selection. Dobzhansky observed that small changes of genetic material within fruit flies could, over time, lead to significant change in the morphology of the fly. Coupled with the theoretical work of Fisher, this gave him the confidence to extrapolate observations of small-scale change, i.e. microevolution, to the evolution of large-scale difference amongst organisms, i.e. macroevolution.[26] Many, like Dobzhansky, began to take on faith that population genetics had the answers to the problem of macroevolution despite the fact that this position rested upon an unproven long-range extrapolation.[27]

The second intellectual giant of the synthesis, Ernst Mayr, had a unique role because he was a naturalist by trade in a synthesis dominated by geneticists and mathematicians. The influence of Mayr's acceptance of the Neo-Darwinian paradigm is hard to overstate, given his standing outside the field of population genetics — acceptance stemming not from allegiance to a mathematical model but rather from observations of actual bird populations, data much more appealing to naturalists and systematists.

Mayr observed that when one examined various bird populations in the wild, different degrees of speciation could be observed. For example, one could find two distinct populations of birds that as a whole had not diverged from one another, another population in which small peripheral isolates had diverged to the point that they would be considered subspecies, as well as everything in between. Mayr argued that if one could see the various stages of speciation in the wild, then speciation must progress at a gradual and continuous rate. Conveniently, his ideas were in accord with the models proposed by Fisher and Haldane. Having corroborated the models of population genetics with his observations of nature, Mayr applied gradual change to the whole of evolution. [28]

Population genetics was established as *the* field of evolutionary study. Disciplines that examined whole organisms, such as paleontology, were marginalized by the reduction of evolution to the interaction between favorable and unfavorable genes. In the view of Neo-Darwinism's critics, relatively simple mathematical models, which could never fully explain something as multi-dimensional as a living organism, were permitted to dictate almost completely our understanding of evolution: "The whole real guts of evolution — which is, how do you come to have horses and tigers, and things — is outside the mathematical theory."[29] Dissent from the mathematically informed synthesis was not taken lightly.

Dissidents

But there was, indeed, dissent. Most of the objectors were dealt with swiftly, marginalized via attacks from the leaders of the modern synthesis. The best example of this can be seen in the career of the German-born geneticist Richard Goldschmidt (1878–1958). Goldschmidt, like Mayr, was wary of pushing mathematics too far in the study of biology, believing that the mathematical models were only useful in so far as they mimicked what happened in nature. Based upon his experience, Goldschmidt had to propose mechanism capable of creating new species. He advocated large-scale or *systemic mutations* for this formidable task, though these did not have to occur in one step.[30] Goldschmidt believed that a new pattern of genomic organization arising from wholesale chromosomal rearrangement would provide the novelty needed to create species rapidly, and he was encouraged by the presence of such drastic mutations in fruit flies that lead to legs growing where antennae should or vice versa. However, Goldschmidt could never verify that such mutants, infamously called "hopeful monsters," actually led to the formation of new species. Moreover, his views were in direct opposition to Fisher's mathematical argument, which demonstrated the likelihood that large mutations would reduce an organism's fitness, most likely killing it. In addition to Goldschmidt, German paleontologist Otto Schindewolf (1896–1971) also advocated rapid transformation of species, although he based his argument primarily on the lack of transitional forms in the fossil record.

To help silence the dissidents, in 1953 H. B. D. Kettlewell (1907–1979) set out to demonstrate the effects of natural selection in the wild, and by implication, confirm Darwin's theory. The story of Kettlewell and his famous experiment has recently been thoroughly documented.[31] He took a particular species of moth, *Biston betularia*, and released large numbers of light and dark colored varieties, then observed where they alighted. By counting the number of survivors in areas with dark and light trees, Kettlewell claimed to show how color affects survival. With his experiment, all that Kettlewell could have hoped to demonstrate was the role of natural selection in optimizing a population for particular environmental conditions — a fact never in dispute in the evolution controversies. Unfortunately Kettlewell's experiment was marred by either fraud or shoddy technique.[32] and he later committed suicide. But regardless, it proved nothing about evolution in the Darwinian sense, because there was no innovation, no creation of new species, and no new information — just a change in relative frequencies of moth colors. Regrettably, the experiment is still often presented as evidence for Darwinian evolution.

VI. Modern Battles Over Evolution: 1960–present

Over the last forty years, despite many successes, the Neo-Darwinian school has increasingly found itself under attack from both scientists and non-scientists alike. By the early 1960s, the materialistic implications of Neo-Darwinism were setting off alarm bells in the evangelical community, which led to a resurgence of Young-Earth Creationism. Most early Creationist efforts tended to be rather amateurish, but over the years quality has grown steadily. With increasing numbers of converts, especially among scientists, some modern Creationist research and analysis is fairly sophisticated, if limited in quantity. Its premises, however are completely rejected by a great majority of scientists.

In addition to continual and mounting criticism from Creationists, Neo-Darwinists have had to contend with prominent scientists who have publicly broken ranks regarding Darwin's theory. This has spawned a new school of thought, the "Meta-Darwinian school." Members of this school have proposed new mechanisms for evolution from the self-organizing qualities of matter to symbiosis. Because the Meta-Darwinists still advocate natural mechanisms, their critiques of the Neo-Darwinian paradigm have held considerable sway amongst the academic community.

To make matters even worse for the Neo-Darwinists, a group of individuals, who were not willing to embrace all the premises of the modern Creationists (particularly the young age of the earth) but were nonetheless dissatisfied with the naturalistic explanations for evolution, started their own movement called the "Intelligent Design school." This movement, aided by a number of publications and think tanks, has quickly become a formidable participant in the debate.

As a result, recent attacks on Neo-Darwinism have come from all sides on a variety of topics and it has caused organizations such as the National Academy of Sciences (NAS) and the American Association for the Advancement of Science (AAAS) to publish position papers in response to these criticisms. These modern critics have taken many old arguments, which the "New Synthesis", in their opinion, failed to address properly, and have reformulated and augmented them with modern data. These critics have also developed some new arguments based on more recent developments such as information theory and the stunning advances in molecular biology. The modern battles center on six key issues:

- The discontinuous nature of the fossil record given the gradual, incremental nature of evolutionary change under Neo-Darwinism

- The extent to which the complexity of biological systems exceeds that which can be created by the processes envisioned by Neo-Darwinism
- The ability of organisms, which represent functionally integrated wholes, to tolerate the changes needed for macroevolution
- The validity of extrapolating from microevolution (with its observable mechanisms) to macroevolution
- The ability of random processes (i.e. random mutations) to create new genetic information
- The age of the earth (Creationists only)

At present, then, there are four schools of thought: (1) the dominant Neo-Darwinian school, with the allegiance of the majority of faculty and major scientific organizations; (2) the Meta-Darwinian school, comprised of scientists who reject the adequacy of the Neo-Darwinian paradigm while still favoring naturalistic explanation for the history of life; (3) the Intelligent Design school, which accepts nearly all of modern science, including the age of the earth and the universe, but which rejects the notion that natural processes can fully account for the complexity and variety of life; and (4) the Young Earth Creationist school, which rejects the idea of an old earth, and consequently any possibility of evolution in the Darwinian sense (while accepting that degenerative change can occur)

References

1. Condensed from chapter 2 of *The Evolution Controversy*, by Thomas Fowler and Daniel Kuebler, Grand Rapids: Baker Academic, 2007.
2. Aristotle *Parts of Animals*, Loeb Classical Library, Translated by A.L. Peck (Cambridge: Harvard University Press, 1937, 1968), p. 101. 645a: 31–37.
3. Benjamin Wiker, *Moral Darwinism: How We Became Hedonists* (Downers Grove, IL, Intervarsity Press, 2002), 63.

4. Steno, Nicolaus, *The Prodromus*, Translated by J.G. Winter (Norwood, Mass.: Norwood Press, 1930), 219.

5. Margaret Espinase, *Robert Hooke* (Berkeley: University of California Press, 1962), 76–77.

6. Georges Cuvier, *Essay on the Theory of the Earth.* Translated by Robert Jameson (New York : Kirk & Mercein, 1818), 99.

7. Georges comte de Buffon, *Natural History*, Translated by J.S. Barr (London: Barr, 1792).

8. Geoffroy St. Hilaire, *Philosophie anatomique.* Vol 1 (Paris: J.B. Bailliere, 1818) 9.

9. University of California Museum of Paleontology, History of Evolutionary Thought, *http://www.ucmp.berkeley.edu/history/lamarck.html* (accessed June 1, 2006).

10. Lamarck JB, *Philosophie zoologique*, trans. H. Elliot (New York: Hafner Publishing: New York, 1963) 119.

11. Thomas Malthus, "A Summary View of the Principle of Population 1830" in *Three Essays on Population* (New York: Mentor Books, 1960), 13.

12. Edward Blyth, "Varieties of Animals," *Magazine of Natural History*, 8 (1835):40–53. Text available in Loren Eiseley, *Darwin and the Mysterious Mr. X* (New York: Harcourt, Brace, Jovanovich, 1979), 97–111. Quote in the text is from page 103.

13. Fred Hoyle, *The Mathematics of Evolution*, (Memphis: Acorn Enterprises LLC, 1999, p. 105.

14. Luther Sunderland, *Darwin's Enigma*, (Green Forest, AR: Master Books, 1998), 18.

15. *Ibid.*, 56–7.

16. Darwin, Origin of Species, 317.

17. Huxley letter

18. Eiseley, *Darwin's Century*, 210.

19. Julian Huxley, *Evolution: The Modern Synthesis* (New York: Harper and Brothers, 1942), 38.

20. JBS Haldane, *The Causes of Evolution* (London: Longmans Green, 1932), 113–114.

21. This is very similar to the tempo proposed by Gould and Eldridge's theory of punctuated equilibrium. See chapter 9 of *The Structure of Evolutionary Theory* for a discussion of their theory.

22. Sewall Wright, "The roles of mutation, inbreeding, crossbreeding and selection in evolution," *Proc. 6th Intl Cong. Genet.* 1932, 358–9.

23. It is important to note, however, that this problem is far more difficult than Wright's comments, and the accompanying figure, suggest. Because the number of ways a genome can vary is quite large, the "space" which must be searched to find suitable peaks is vast—so vast that the trial-and-error method Wright proposes would only be able to search a miniscule fraction of it even over evolutionary time spans.

24. See www.evolutionprimer.net

25. This is not atypical of the way science proceeds, of course. Scientific theories are almost always the product of a long and difficult road, and as such are defended tenaciously as some form of absolute or near absolute truth. But at the same time, they are subjected to ruthless scrutiny and relentless testing, so that any empirical deviation from theory, however small can be found, and its impact on the theory assessed.

26. Theodosius Dobzhansky, *Genetics and the Origin of Species* 3rd edn. (New York: Columbia University Press, 1951), 16–7.

27. Dobzhansky, Genetics and the Origin of Species, 12.

28. Ernst Mayr, Populations, Species, and Evolution: An Abridgement of Animal Species and Evolution (Cambridge: Harvard University Press, 1970), 351.

29. C. Waddington, "Discussion [of paper by Murray Eden]", *Mathematical Challenges to the Neodarwinian Interpretation of Evolution*, ed. P.S. Moorehead and M.M. Kaplan (Philadelphia: Wistar Institute Press, 1967), 14.

30. Goldschmidt, Richard. *The Material Basis of Evolution* (New Haven: Yale University Press, 1982), 206.

31. Judith Hooper, Judith, *Of Moths and Men: An Evolutionary Tale* (New York: Norton, 2002), xvii.

32. Hooper, *Of Moths and Men*, 262.

Charles Darwin (12 February 1809–19 April 1882)

9. Life's Intelligible Design

Manuel Alfonseca

Creationist groups in the United States have updated their message. Their one-century old attempt to eliminate evolution from school curricula, which gave rise in 1925 to the famous trial of the state of Tennessee against John Scopes (and in 1960 to Stanley Kramer's film *Inherit the wind,* based on that trial) has gone through three different and subsequent steps:

1. They first tried to attain state legislation forbidding evolution to be taught in schools. They were successful in the Arkansas Statute of 1928, which was revoked by the U.S. Supreme Court in 1968.

2. They next attempted to reach *balanced treatment,* i.e. forcing schools to dedicate the same time to explaining evolutionary theory and *creation science* (whatever that is), based on a literal interpretation of the first chapter of Genesis. They were successful in 1981 in the state of Louisiana, but this decision was also revoked by the U.S. Supreme Court in 1987.

3. The third attempt is *intelligent design,* which in a few words can be summarized as *the assertion that God directs and controls evolution in detail.* In other words, that some of the evolutionary phenomena are not explainable by the free action of chance, the environment and natural selection, and require the direct intervention of a divine being. In their view, this *scientific theory* should be taught on equal grounds with standard evolutionary theory in the text books on natural sciences used in the first and second levels of education.

Intelligent design, as defined by creationists, asserts that there is something in living beings which is impossible to explain as an effect of chance. To prove it, they provide several cases, based on the existence of very complex organs, such as the eye, or rotating flagella in bacteria, or on complicated behaviors, such as wasps which paralyze spiders by injecting venom in each of their nervous ganglia. These arguments are usually presented as though they were new and unanswerable, while some of them actually go back to the time of Darwin himself, or to Henry Bergson[1] and his theory of *l'elan vital* at the beginnings of the twentieth century, and have been rejected long ago by evolutionary scientists[2]. For instance, it is a mistake to assume that the evolutionary process towards the apparition of a complex eye must be the result of a gradual evolution through intermediate species with partial or imperfect visual organs, unable to perform usefully. In actual fact, the evolution of the eye may have taken place as a result of a few mutations, each of them indifferent or with selective value.

This attempt has induced the indignation of many scientists, who accuse the proponents of *intelligent design* of trying to slip a purely philosophical or religious theory as a scientific alternative to the theory of evolution. The problem is, that some men of science who defend the theory of evolution get a step further and fall into the same *sin* with which they charge their opponents, presenting philosophical speculations and dogmatic statements as though they were verifiable scientific theories. Philosophical connotations are not scientific, both if one says, with some believers, that *there is an intelligent design behind everything,* as in the opposite stance, that *everything is the result of mere chance.*

It is impossible to reach an agreement in this problem. Let us assume that there is something in living beings which we cannot explain just now as an effect of chance. In that case, an atheist scientist can always say that there exists some still unknown cause which, when discovered, will explain satisfactorily the pending question, and the theist who sees in it a mark of the hand of God will be accused of believing in a God of the gaps. On the other hand, everything we know about living beings may very well be compatible with the action of apparent chance. However, even in this case, the metaphysical hypothesis of intelligent design would not be automatically excluded, for God may have included chance among the tools he has used to create the universe. Or are we going to deny God the capability of using mechanisms which we ourselves do use?

Neither intelligent design nor evolution by pure chance are scientific theories, for it is impossible to prove them false. Both are metaphysical theories and should be presented in that light. Text books on natural science must not present intelligent design as an alternative to the scientific theory of evolution, because it is not. But neither should they assert that science has proved that *God does not exist*, or that *the universe and life are the result of pure chance*, because these two assertions are simply false: science cannot prove any of those things.

The scientific theory of evolution

As every scientific theory, evolution is, and always will be, a provisional theory, but after one century and a half, it is now quite well checked by arguments coming from very different fields of science, such as:

- Comparative anatomy
- Embryology
- Paleontology
- Biogeography
- Molecular biology

It is thus very improbable that the theory of evolution may be knocked down by new advancements of science, further than being subject to modifications of detail, as always happens with every scientific theory.

The first thing to be straightened out is what we understand by the *scientific theory of evolution*. The use of the word *scientific* implies that this theory has to do with

1. verifiable facts, and
2. hypothesis that explain them, which should always be subject to the possibility of proving that they are false[3].

In this context, the theory of evolution is based on the verifiable observation that species change with time, and analyzes the mechanisms by means of which this can happen: mutations, DNA, natural selection, and so forth.

In its current form, evolution depends on the interaction of three main factors[4]:

1. The genetic variability of living beings, which is stored mainly in DNA, is inherited from one generation to another following (usually) Mendel laws, and is modified randomly as a consequence of the action of cosmic rays and other radiations, the exposition to certain chemical substances, genetic recombination between chromosomes during sexual reproduction, and other alterations that can take place during cellular reproduction.
2. The environment, which also acts and changes randomly, modifying the conditions in which living beings must live and reproduce.
3. Natural selection: the fact that the individuals best adapted to the environment have more chance of surviving and reproducing, in this way transmitting their genes to the next generation.

A fourth element should be mentioned, although it rarely is: the universal constants and the basic laws of nature, which, in our current scientific view, are general relativity and quantum mechanics. These constants and laws seem to be finely tuned to make life possible. The problem of fine tuning provides a strong indication (but not a scientific proof) toward the possibility of design.

Research on Artificial Life

There are inklings indicating that evolution is compatible with intelligent design. In a recent branch of computer science (evolutionary programming), procedures inspired on biological evolution are applied to program building. In one of its subdivisions, *artificial life*, evolutionary programming is used to develop agents which remind the behavior of living beings. One example is the simulation of ant colonies, which throws light on the behavior of swarms of beings who act together and let us formulate hypotheses about the emergence of higher order entities: multi-cellular organisms, or insect and human societies[5] (see below). Experiments on artificial life can also be used to simulate artificial ecologies, or to study the transmission of language among groups of human beings, simulated as drastically simplified agents, among many other applications.

The results of the experiments on artificial life depend on four factors:

1. *Random variations* (mutations and recombination) in the *genomes* of the simulated *living beings*, which actually are not really random, but pseudo-random, since appropriate algorithms are used to generate them.
2. The *environment*, which can also change in a pseudo-random way.

3. Natural selection, that favors the survival of those genomes more adapted to the environment through the computation of a fitness function.
4. The *basic rules of the game*: the set of instructions that the *living beings* in the experiments can combine to initiate their evolution, and which affect their behavior.

It is easy to see that experiments on artificial life parallel biological evolution. Obviously, however, seen from our point of view, an experiment in artificial life is a case of intelligent design by the programmer. In these experiments, agents interact under the control of algorithms which use series of pseudo-random numbers, i.e. under the control of pseudo-chance. If some day we were able to produce intelligent agents in our simulations, these agents would not be able to deduce our own existence by experimentation, for how could they experiment on us, who are outside their world? However, there is another possibility: would they be capable, by analyzing in some way the randomness of their world, to detect that, rather than random, their *chance* is pseudo-random, thus discovering a proof of our existence?

There is a mathematical theorem[6] that proves that, given a sufficiently complex succession of integers of unknown origin, it is impossible to distinguish whether that succession has originated randomly or pseudo-randomly. Therefore, neither my hypothetical artificial intelligent beings would be able to detect the pseudo-randomness of the *chance* we introduce in their evolution, nor we ourselves will ever be able to distinguish between those two possibilities in our own universe: chance evolution or *providential evolution* (I prefer this name to *intelligent design*). In other words, perhaps the apparent chance of our universe is pseudo-random for God.

If my hypothetical beings came to the conclusion that their world exists by mere chance, they would be wrong, because we know that their world has been designed by us.

Our role with respect to them would be (saving the obvious differences) parallel to that of God with respect to us. In the hypothetical artificial life environment I have described, the atheistic metaphysical argument which states that the universe has not been created by anybody and has developed by itself as a purely random process would be false, but our agents would be unable to prove it or disprove it. It is thus evident that the same atheistic argument, applied to our universe, may not be valid, but we cannot prove it either way. Therefore, this question is unscientific.

How can God act in the world?

Let us then make metaphysics. Assuming that God exists, and that the universe has been the object of intelligent design, how can God act, in which way can God interact with the cosmos, lead its development according to His design?

Obviously, God could manipulate the universe by skipping the laws He has given it, i.e. through miraculous actions. However, from the study of nature and history, it seems that this type of divine action, if it happens, is extremely rare. God hides from us with an exquisite care. He does not seem to wish that His existence can be proved unassailably, perhaps because otherwise He would be forcing our will and cancelling our freedom.

Can God manipulate the universe, modify its evolution, without our knowing it? This would be a special type of divine action, which is usually called *providence*. How could He do it? I will contend that God, in his providence, can interact with the world in three different ways to carry out his intelligent design:

- Upon the deterministic side of the world (represented by

general relativity), God can act by manipulating the initial conditions of the universe. For God, such conditions would not be affected by chaos theory, as He would be capable of setting them to an infinite number of decimal figures. We must assume that the actual value of real numbers (which we cannot know) is not outside God's intellect, since He invented them. On the other side, Heisenberg's uncertainty principle is a restriction applicable to us, who are within the universe, not to the creator, who is outside. Given that God is also outside time, He would be able to use his global knowledge of the cosmos to settle the initial conditions in such a way that certain events may take place in any subsequent future.

For instance, we believe that the extinction of dinosaurs took place sixty five million years ago as a consequence of the impact of a heavenly body (an asteroid or a comet) against the Earth. After the disappearance of dinosaurs, the mammals, up to then reduced to a small size, found an open field to evolve and invaded all the ecological niches occupied by large animals, which suddenly had become free. This gave rise to the evolutionary branch which led to man. We can imagine that, in an intelligent design leading to the appearance of an intelligent species on Earth, God could have planned that impact since the initial conditions given to the universe at the beginning. Phenomena of this kind happened more than once during the history of the Earth and would have made it possible to apply important modifications to the evolutionary process.

- Upon the indeterministic side of the world (represented by quantum mechanics), God could act by manipulating the random subatomic quantum effects, while carefully maintaining statistical equilibrium, in such a way that those effects were for us indistinguishable from chance. Again, although this operation would be inconceivable for us, we must assume that it would not be out of the reach of an intellect capable of designing and creating an universe as complex as ours, in the same way (saving the difference) as I am able to manipulate at will all the features of the computer programs under my artificial life experiments.

- Once man appeared, equipped with free will, God can use conscious human beings as a bridgehead to enter directly (but undetectably) in the material world. This could be done by suggesting a concrete human being to perform a given action, usually, but not always, through their conscience or sense of duty. In this case, however, against the two previous cases, God runs the risk of failure. In other words, we can fail him, for He never thwarts our freedom.

What is life?

One interesting question concerns the definition of life. In other words, is there any way to be sure that life is present or not?

It is evident that some beings possess life and some do not. We, all the plants and the animals, are alive; stones, distilled water and carbon dioxide, are not. When Antony van Leeuwenhoek discovered microorganisms, nobody doubted that they were alive. The problem came with the viruses, which are much smaller than cells, being made of a single DNA or RNA molecule inside a protein capsule. Initially they were considered mere poisonous chemical substances (thus their name). Then they were thought to be tiny living cells, invisible to the optic microscope. In 1935, Wendell Meredith Stanley proved that the virus causing the tobacco mosaic disease can crystallize, the same as many chemical substances. For the second time it appeared that viruses were not alive. Now, however, they are usually considered as parasitic cells, reduced to the minimum expression. Crystallizing does not have to be incompatible with life.

The ability to reproduce is, for the time being, exclusive of living beings. Perhaps we should consider this function as the touchstone defining life, what distinguishes living from inert beings. This criterion would classify viruses as alive. Another common feature for all living beings (at least those that we know on the Earth) is the fact that they use nucleic acids (DNA and RNA) to code their genetic information.

There are isolated RNA molecules, called *viroids,* with just a few hundred nucleotids and without a capsule, who are able to reproduce parasitically inside cells and cause diseases to plants. This opens the question whether nucleic acid molecules, by themselves, should be considered living beings. Biologists do not agree on this issue. If this were accepted, the origin of life would have happened before we thought[7], and life could be classified in four different levels:

- First level: isolated nucleic acids similar to viroids.
- Second level: prokaryotic cells (bacteria and Archaea). Each prokaryotic cell contains several nucleic acids working together for the common welfare of the prokaryotic cell.
- Third level: eukaryotic cells. Each eukaryotic cell contains many prokaryotic cells (mitochondria and/or chloroplasts) which work together for the common welfare of the whole cell.
- Fourth level: multi-cellular beings, each of which is made of many (sometimes trillions) eukaryotic cells, all working together for the welfare of the whole organism.

This process seems to be recursive, for we do have now some examples of incipient living beings of the fifth level, i.e. groups of organisms who work together for the welfare of the whole super-organism: coral reefs, anthills and beehives can almost be considered as independent living beings, whose cells are polyps, ants and bees. To a certain extent, human society is on the way to become something similar (but, we hope, better).

What is man?

After the consolidation of evolutionary theories, many biologists tend to assert that man is an animal like any other, just one between millions of species of living beings. According to them, it is impossible to set criteria that help us decide if one species is more advanced or more important than any other.

Is this true? I think not. It seems evident to me that those criteria do exist, that we should not be denied the ability to compare and judge, two capabilities that have made all our technological advances possible. I'll mention just two of those criteria:

The origin of life, about 4000 million years ago, did not have an immediately observable impact on the physical aspect of the earth, apart from some changes in the sea water hue, or the apparition of reefs of cyanobacteria. Nonetheless, the action of life on Earth continued slowly and culminated, about 1000 million years ago, in a new composition of the atmosphere, with 20 percent oxygen, which made respiration possible. This was a truly global change of the Earth provoked by the joint action of the first three levels of life (see above).

With multi-cellular living beings, the physical appearance of the Earth changed deeply: the dominant color of continents became green, rather than brown. Of the three kingdoms at the fourth level of life, plants produced the largest impact: fungi and animals are practically imperceptible from outer space.

In the last centuries, the situation has changed: for good or evil, the human species, just by itself, has modified the appearance of our planet. The surface of the tropical forests is getting smaller; continents are crisscrossed by huge networks of roads and railroads; the night sky of our cities is full of light; seen from space, the night side of the Earth is now lighted in many places; a large proportion of living species is in danger; holes in the ozonosphere turn up; and, for the first time in its history, the Earth has become a strong emitter of low frequency electromagnetic waves (radio and microwaves), which make our existence detectable by hypothetical extraterrestrial intelligences. A single species has done this in a terrifically short time, compared to the history of the Earth.

On the other hand, in very recent years, the human species has come to manage a huge amount of information, over 10^{18} (one quintillion) bits, a number which is still growing. Compare this with the maximum information computed by any other species, from bacteria to chimpanzees: between one million and 200 million bits, seven to twelve orders of magnitude below ours (see Figure 1 and Table 1). The information at our disposal may already be higher than the total amount of information accumulated by the hundred million species of living beings considered to have existed from the origin of life to our time, assuming that it makes any sense to add it all together.

Is man a species like any other? Not at all. Man is unique. In proportion to its body, its brain is larger than that of any other living species, able to store about 10 trillion bits (10 Tbit): over one thousand times more than its genome; 50 times more than most mammals. It appears that, with the apparition of man, life crossed a critical point which, for the first time in history, allowed one individual to reach such levels of information.

Five thousand years ago, with the invention of writing, man has crossed a second critical point, becoming the only species that possesses a third way to store information, a memory outside its body. With the invention of computers and the Internet, that information is now available to everybody and still grows. Every human being can access a total information ten million times greater than that contained in

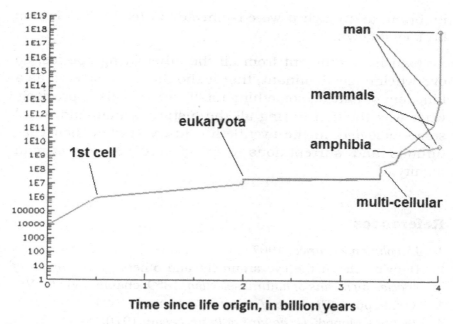

Figure 1. Information available to different kinds of living beings. Green: genetic information. Red: Total information (genetic + neural + cultural). Only man can use cultural information.

Table 1. Amount of information at the disposal of different kinds of living beings.

Living being	Genetic information	Neural information	Cultural information
Virus	10–50 kbit		
Bacteria	1–10 Mbit		
One-cell eukaryote	25 Mbit		
Nematode	200 Mbit	5 kbit	
Amphibia	2 Gbit	500 Mbit	
Reptiles	3 Gbit	5 Gbit	
Mammals	5–6 Gbit	200 Gbit	
Man	6 Gbit	10 Tbit (10^{13} bit)	10^{18} bit

its brain, as though it were connected to ten million brains besides its own.

Man is so different from all the other living species, so overwhelmingly dominant, that it should be considered as a kingdom of nature, something totally apart. This is precisely what was thought during all the history of mankind, until some biologists in the twentieth century started their continuous and surreptitious work of undermining human dignity.

References

1. *L'evolution creatrice*, 1907.
2. Darwin refuted the eye argument and others in *The origin of species by means of natural selection*, 1859, chapters VI and VII.
3. Karl Popper, *The logic of scientific discovery*, 1934.
4. Jacques Monod, *Le hasard et la nécessité*, 1970.
5. M.Alfonseca, J. de Lara: *Two level evolution of foraging agent communities*, BioSystems, 66:1–2, pp. 21–30, Junio-Julio 2002.
6. Gregory Chaitin, *Randomness and Mathematical Proof*, Scientific American 232, No. 5 (May 1975), pp. 47–52, http://www.cs.auckland.ac.nz/~chaitin/sciamer.html.
7. The origin of life raises a historic, rather than a scientific problem, since it only happened once.

Gregor Mendel (July 20, 1822–January 6, 1884)

10. What are the Contributions of Genetics to the Understanding of Life?

Nicolas Jouve

Genetics is a branch of Biology born at the beginning of the 20th century, after the rediscovery and appreciation of the work carried some decades before by the Augustinian monk Gregor Johan Mendel (1822–1884) born in Heizendorf (Chequia). Mendel had performed crossbreeding among varieties of the pea *Pisum sativum* in the garden of Saint Thomas's Abbey in Brno and had carried out careful analyses of how the characters which made them different were transmitted to the descendants. In his explanations, Mendel used a simple symbology and he was able of explaining the inheritance along several generations. The results of his work were presented at two sessions of the Society of Natural Sciences of Brno and were published a year later in the Bulletin of the Society under the title *Versuche über Pflanzen-Hybriden*. In this publication were registered the foundations of the inheritance and the hypothetic deductive model of what later on was called "Mendelism". However, the publication did not get the deserved attention and remained unknown till 1900. While Mendel did not get explicit recognition of his work when he was alive, he said in the Monastery garden, among his plantations of fuzzy plants, peas and *Cirsium* that "my time has not come yet."

One reason to explain the period in which Mendel's work was ignored lies in the prevailing conditions in Natural Science at the time, coincident with the maximum diffusion and popularity of Charles Darwin's doctrine (1809–1882) of evolution. This English author published a book in 1868 on "the variations of animals and plants for their domestication,"[1] in which no mention is made of Mendel's experiments. In the second edition of Darwin's principal work *On the Origin of the Species by means of Natural Selection*, published in 1876[2], he said: "Inheritance's laws are for the most part unknown. Nobody knows why the same peculiarity is sometimes transmitted and sometimes not, and why the child is sometimes alike the grandfather, sometimes the grandmother and ever a previous ancestor." Mendel had solved those open questions at least ten years before.

In the year 1899, the English biologist William Bateson (1861–1926) presented a communication in the International Congress organized by the Royal Horticultural Society of London in which he said: "What is needed to know first is what happens when one variety is crossed with its nearest relative. If the result is to have a scientific value it is almost indispensable to examine statistically the offspring. Note must be taken of the number of descendants which are alike one or the other progenitor. If the parents differ by various characters, the offspring must be studied statistically with respect to each character separately. What matters is to know numerical results approximately." What Bateson was expounding was nothing else by the conditions Mendel had fulfilled scrupulously 34 years before in his analysis of inheritance.

The following year, three researchers, the German Carl Correns (1864–1933), the Dutch Hugo de Vries (1848–1935) and the Austrian Erich Tschermak (1871–1962), all of them working on the inheritance of characters in different organisms, came to know the pioneering work of Mendel and they accepted the priority of his work.

When Bateson knew the works of Correns, De Vries and Tschermak, and through them those of Mendel himself, he was so much impressed that he exposed them at the Royal Horticultural Society of London, becoming immediately the most enthusiastic propagandist of Mendelism. To Bateson is due the name of Genetics of the new born science, as well as the terms allele or allelomorph, homozygosis and heterozygosis, to designate each alternative element of the genes and their pure or mixed condition in the genotype of the individuals, respectively. The term gene, referred to the "hereditary factors" of Mendel, was adopted in 1903 by the Danish investigator Johansen (1857–1927), who introduced also the terms genotype and phenotype.

At the same time, a series of discoveries by various cytologists were to have a very positive influence for the consolidation of Mendelism. First, the German biologist August Weismann (1834–1914) who, in one of his essays published between 1883 and 1885, established the "Theory of the continuity of the Germinal Plasma" according to which the hereditary material passes intact through the germinal line from one generation to the next, against the theory of the acquired characters of the French evolutionist Jean Baptiste Lamarck (1744–1829). Weismann stressed the separation between germinal and somatic cells and proposed that the chromosomes of the sexual cells were the carriers of the hereditary characters. This way it was originated the idea of correlating the chromosomic behaviour which cytologists had described through the observation of the mitosis and the meiosis, with the behaviour of Mendelian factor. The observation of greatest significance was made by the Belgian cytologist Edouard Van Beneden (1946–1910) who in 1883 discovered the mechanisms of the meiosis and of the karyogamy. After those works, those of the German Theodor Bovery (1962–1915), and the American Edmund Wilson (1856–1939) and Walter Sutton (1877–1916), who established the "chromosomic theory of the inheritance" in 1902, through which it was affirmed that the chromosomes are the physical see where the chromosomes are located.

Since then, Genetics has done nothing else but adding knowledge and scattering a body of doctrine in a series of specialities in interaction with other branches of Biology, more set up on the study of the properties and functions of the living beings than the traditional classification of the species or the description of the constitutive elements. In the following table the main landmarks in the development of Genetics are registered.

Main Landmarks In The History Of Genetics

1858 — **Charles Darwin and Alfred Wallace** made the joint presentation of the theory of biological evolution by means of natural selection before the Linneal Society of London.

1865 — **Gregor Johan Mendel**, made crossbreeding in peas and did a meticulous analysis of inheritance.

1869 — **Friedrich Miescher** discovered a substance, "nucleine," of acid character in the nucleus of the cell. Later it became known a DNA.

1883 — **August Weismann**, issued in the theory of the continuity of the Geminal Plasma.

1900 — **Hugo de Vries, Carl Correns and Erich Tschermak**, rediscovered Mendel's laws. Valued the work of the Augustinian monk and in this way the science of the inheritance was born.

1902 — **Walter S. Sutton and Theodor Bovery** put forward the "Chromosomic Theory of Inheritance."

1906 — **William Bateson** made known Mendel's discoveries and coined the term "Genetics" to designate the science of the inheritance.

1909 — **Alfred Garrod** discovered a human illness, the alcaptonury which is due to a genetic mistake. This way was born the idea of what was later called "congenital mistakes of metabolism."

(*Continued*)

(Continued)

1926 — **Thomas Hunt Morgan**, introduced the vinegar fly, Drosophila melanogaster in the genetic studies and published "The Theory of the Gene." Nobel Prize 1933.

1928 — **Frederick Griffith**, was able to transform permanently non virulent lineages of pneumocoques.

1941 — **George Beadle and Edward Tatum** proposed the hypothesis: "one gene-one enzyme": the genes determine the synthesis of the enzymes. Nobel Prize 1948.

1944 — **Oswald T. Avery, Colin M. McLeod and Maclyn** MacCarthy, demonstrated that the DNA, not the proteins constitute the substance of the genes.

1946 — **Hermann Muller**. Worked in the induction of mutations caused by X-rays. Nobel Prize 1946.

1948 — **Joshua Lederberg**. Analyzed genetic recombination in bacteriae. Nobel Prize 1948.

1952 — **Alfred Hershey and Martha Chare**. Added new proofs in favour of the fact that DNA is the molecule of inheritance.

1953 — **Frederich Sanger**. Developed the techniques for the sequentiation of proteins and sequentiated the firs protein, human insulin. Nobel Prize 1958.

1953 — **James Watson and Francis Crick**. Discovered the "double helix" structure of DNA. **Rosalind Franklin** must be also mentioned. She achieved magnificent X-ray diffraction patterns which allowed unveiling the DNA structure. Watson, Crick and **Maurice Wilkings** received the Nobel Prize in 1962.

1955 — **Barbara McClintock**. Discovered the phenomenon of transposition (mobile elements) in corn. With some delay received the Nobel Prize in 1983.

(Continued)

(Continued)

1958 — **Edward Tatum and George Beadle**. Extended the understanding of the congenital errors of metabolism and demonstrated the theory of "one gene-one enzyme."

1959 — **Severo Ochoa and Arthur Kornberg**. Discovered polymerization enzymes for RNA and DNA. Nobel Prize 1959.

1961 — **Francois Jacob, Jacques Monod and Andre Lwoff**. Discovered the regulation systems of the genic expression in microorganisms. Nobel Prize 1965.

1961–64 — **Robert W. Holley, Godino Khorana and Marshall Niremberg**. Unveiled the "genetic code" and its role in the synthesis of the protein. Nobel Prize 1968.

1962 — **Max Delbrück, Salvadore Luria and Alfred Hershey**. They made important contributions to the understanding of the biological cycle of the viruses and the preadaptative character of mutations. Nobel Prize 1969.

1968 — **Werner Arber**. Identified the restriction enzyme, making it possible genetic engineering.

1969 — **Jonathan Beckwith** and collaborators were able of insulating for the first time one gene in the bacteria Escherichia coli.

1970 — **Norman Borlaugh**. Creator of the "green revolution" with his genetically improved plants. Nobel Prize 1970 (for Peace).

1970 — **Renato Dulbecco, Howard Temin and David Baltimore**. Discovered the enzyme "reverse transcriptase" and the interaction between tumoral viruses and the cell genetic material. Nobel Prize 1975.

1977 — **Frederick Sanger and Andrew R. Carlson**, in Cambridge (U.K.) and Allam Maxam and Walter Gilbert, in Harvard (U.S.A.) developed the techniques of sequenciation of DNA. F. Sanger received his second Nobel Prize in 1980, shared with W. Gilbert.

(Continued)

(Continued)

1978 — **Walter Gilbert, Richard J. Roberts and Phillip Sharp**. Discovered the structure of "introns" and "extrons" of the genes. R. Roberts and P. Sharp received the Nobel Prize in 1993.

1978 — **Werner Arber, Daniel Nathans and Hamilton Smith** discovered the restriction endonucleases and opened the field of their application in Molecular Biology. Nobel Prize 1978.

1980 — **Paul Berg**. Made important contributions in recombinant DNA. Nobel Prize 1980.

1980 — **Sidney Altman and Thomas R. Cech**. Discovered the catalytic properties of RNA (ribozimes). Nobel Prize 1989.

1984 — **James Watson, Renato Dulbecco**, together with other investigators in U.S.A. promoted the idea of the Human Genome Project.

1985 — The genome of several viruses of bacteriae, MS2 ΦX184, T7 and λ had been deciphered completely, as well as that of mitochondrial DNA and the animal viruses SV40 Epstein — Barr.

1988 — Project Human Genome is approved in U.S.A.

1989 — **Michael Bishop and Harold Varmus** received the Nobel Prize in Medicine for their discovery of the cellular origin of the retroviral oncogenes.

1990 — Project Human Genome is underway, including other species.

1993 — **Kary Mullis** developed and published an important technique in Molecular Biology, the Amplification in the Polymerase Chain. Nobel Prize 1993 (for Chemistry).

1995 — **Edward Lewis, C. Nüsslein–Volhard and Erick F. Wieschaus**. Important discovery on the genetic regulation of the early development of the embryos. Nobel Prize 1995.

(Continued)

(Continued)

1996 — The complete genome of the yeast *Saccharomyces cerevisiae* is published (6.300 genes in 120,5 Mb).

1997 — The complete sequenciation of bacteria *Escherichia coli* is culminated.

1992–99 — The techniques of automatic sequenciation are developed and improved in its efficiency.

1999 — The sequenciation of human chromosome 22 is completed and published.

2000 — The sequenciation of human chromosome 21 is completed and published.

2001 — The "work draft" of the human genome is completed.

2002 — **Sydney Brenner, John Sulston and H. Horowitz.** Contribute to the understanding of the genetic regulation of the development of organs and the programmed cellular death. Nobel Prize 2002.

2003 — In the month of April in synchrony with the 50th anniversary of the discovery of the double helix, **Francis Collins**, director of the public consortium of the Project Human Genome, announces its culmination after completing the information lacking in the draft.

2005 — The project of the chimpanzee genome **Pan troglodytes** is culminated.

2011 — The project of the orangutan genome **Pongo pygmaeus** is culminated.

2012 — The project of the gorilla genome **Gorilla gorilla** is culminated.

After the *Chromosomic Theory of the Inheritance* it was very useful the contribution of the American Thomas Hunt Morgan (1866–1945) who introduced in Genetics the fruit fly *Drosophila melanogaster*. This organism, because its easy manipulation in the laboratory and its genetic simplicity propitiated the experimental demonstration needed for the Chromosomic Theory of the Inheritance and opened the door for the establishment of the first genetic maps.

However, no doubt, the unifying element which pushed forward Genetics and determined its catalytic role in Biology was the understanding of the molecular organization of DNA, the clarification of the concept of gene and its mechanisms of transmission, expression and mutation. The gene constitutes the essential component of the studies in Genetics which have allowed the understanding of numerous problems in Biology, leading to multiple biotechnological applications. In fact, the science of Genetics, because its methodology and because it reaches all the levels of organization of living beings — molecules, cells, individuals, families, populations, species and taxa above the species has contributed decisively to solve the most crucial questions in Biology. In what follows we will see how were successfully answered these questions.

What is the chemical nature of the genes?

Around 1869 Friedrich Miescher (1844–1895), a Swiss biochemist, had succeeded in synthesizing a substance of acid characteristics from the nuclei of human lymphocytes, birds eggs and spermatozoids of several species. He never came to realize the transcendence of his finding. The substance in question was named "nucleine". At the time Miescher was studying the substance of the cellular nucleus, Weismann was advancing the idea that the substance responsible for the genetic information should have a chemical nature and a molecular structure capable of explaining such and important biological role.

From 1910 on the works of Morgan and his school had been instrumental in setting down the idea that the genes are aligned in the chromosomes in spite of the fact that around 1920 their biochemical nature was unknown. For some time, geneticists believed that the molecular candidates to make up the substance of the genes, the proteins, were the ones which fitted best for that important biological role. At that time it was thought that the nucleic acids were monotone and uniform macromolecules.

The rest of the genetic role of DNA begun to be understood when the purification of fragments of these macromolecules was achieved and when it was seen that they had the capacity of modifying genetically receptive bacteria, in other words, of conferring new hereditary characters when they were absorbed in the cultivated medium. This phenomenon, called "transformation," was discovered by the British investigator Frederick Griffith (1979–1941) in 1928, and its biological significance was understood 16 years later. Griffith was studying septicaemia in mice, a sickness caused by pathogenic bacteria lineages of *Diplococcus pneumoniae*.

Griffith was handling two types of lineages, one pathogenic (S) and one non-pathogenic (R). In his experiment, Griffith injected mice with bacteria R and the animals did not die. He injected mice with bacteria S killed by heat and obtained the same result. At last he injected mice with a mixture of bacteria R and bacteria S treated with heat and the animals died of septicaemia and also he saw that bacteria S were extracted form their blood. Griffith's conclusion was that there should be a "transforming principle" that passed bacteria S treated with heat to bacteria R, converting them from non-pathogenic (R) to pathogenic (S). However, he did not approach the study of the transforming principle.

In 1944 the unknown was unveiled thanks to an elegant experiment developed by the American researchers Oswald Avery, Colin MacLeod and Maclyn McCarthy, by means of which it was demonstrated that the DNA and not the proteins were the constitutive substance of the genes. The demonstration counted in "transforming" one bacterian stock of non-nocive neumocoques for the mice in pathogen through the absorption and incorporation to its genome of DNA coming from outside. Since then it became known that the DNA is the "molecule of life" and due to that the discovery signalled the birth of the new era of Molecular Biology and Genetics.

The next step was obliged: How the DNA molecule was configured to satisfy its transcendental biologic role? In 1953 the American James Watson and the British Francis Crick[3] published their conclusions on the structure of the double helix, deduced from the interpretation of the X-ray diffraction pattern of properly aligned molecules obtained by the physicists Rosalin Franklin and Maurice Wilkins. As it had been forecasted by Weisman, the DNA molecular structure presented a perfect design to explain the role of conserving, spelling and allowing the modification of genetic information.

The molecule is really composed of two polymers making up chains of alternating units of desoxiribose sugar and phosphate which turn around in parallel about each other (double helix) and interconnect with each other by means of paired molecules acting as a bridge, called nucleotide bases, of four different types (Adenine, Guanine, Thymine and Cytosine). Each base of one of the chains joins a desoxiribose sugar and projects itself towards the interior of the molecule, pairing with one base of the molecule of the other chain by means of hydrogen bonds. The pairings take place always between an Adenine and a Thymine or between a Guanine and a Cytosine. These internal bonds are commonly known as the stepping elements in the double helix and the determinant factors of the necessary variability for the molecule to act as a carrier of information. In this way, if we observe a certain span of the molecule we see a written language made up of four symbols: A–T, T–A, C–G and G–C. A gene is really a given spans of a sequence of several hundreds of thousands of pairs of bases.

The structure made up of two molecules with a sequence of complementary bases mutually dependent of each other explains how replication can take place, in other words, how two molecules of DNA with the same information can be obtained from the original one. At a given moment the double helix can open up itself like a zipper in such a way that once apart each one of the two filaments of DNA acts as a mold for the synthesis of the complementary filament. It suffices that in front of each base of the old chain locates itself the complementary base of the new molecule.

In what consist the mutations?

The mutations are changes in genetic information which can have an influence on all levels of organization, from the nucleotide bases of the DNA to the complicated chromosomes. In the case of those mutations called point mutations, what undergo variations are the nucleotide bases of the DNA, which could affect to the proper information of a given gene or could take place in extragenic regions. In the lower organisms, like the prokaryotes, the DNA of the chromosomes is immersed in the cytoplasm and free from other molecules. In the so called eukaryotes the chromosomes are organized like packs of DNA in interaction with basic proteins. In this last case to the mutations of changes in the nucleotide bases can be added other changes which can affect the structure and the number of chromosomes in an organism.

The mutations can take place as spontaneous accidents, during the proper replication of the molecules, or during the cellular division, or can be induced by means of mutagenic agents. In any case, many mutations are repaired "a posteriori." To that end the organisms posses diverse reparation molecular systems in which diverse enzymes intervene.

It is important to notice that the mutations are natural events and have a low frequency of occurrence. That frequency is higher in superior living beings in which a gene can mutate with a frequency of 10^{-4} to 10^{-6} per gamete, than it is in the microorganisms, with mutation rates of the order of 10^{-8} to 10^{-9} per cellular generation. These rates are low enough to keep the general characteristics of each species and high enough to contribute to maintain a genetic diversity among the individuals of each species.

How do genes express themselves?

In 1958 Francis Crick coined the expression "Central Dogma of Molecular Biology" to mean the set of molecular mechanisms through which the genetic information enclosed in the DNA makes copies of itself through "replication" or uses it to produce the synthesis of other molecules different from the DNA. In the later case the information expresses itself in two successive steps. The first named "transcription," consists in the synthesis of a molecular intermediary of the DNA, the so called messenger RNA, or m–RNA. The second is the "translation," which involves the synthesis of the aminoacids chain required to build up the proteins following the instructions of the message contained in the m–RNA. It must be recalled that the proteins are chains of variable length made up of aminoacids, and they constitute the final expression of the genes. The first step to decipher the genetic information was given in 1941, when George Beadle and Edward Tatum demonstrated that the proteins are the molecular objectives of the information of the genes.

This is explained by a classic principle in genetics, "one gene–one protein," or "one gene–one enzyme." The proteins are the most important molecules of the living beings because they form part of the structure of the cells, the tissues and the organs, or behave as enzymes responsible for the materialization of the reactions of the metabolism.

The advances in Molecular Biology of the following decades did show the irreversibility of the transfer of information from the gene to the proteins and that to this end it is necessary the existence of a codification system, a "genetic code" which may allow the interpretation of the DNA language, made up of an alphabet with only four letters, to ensemble the aminoacids of the proteins. This operation takes place acting as intermediary a molecule of m-RNA and takes place acting as support certain cytoplasmatic organelles denominated ribosomes.

Soon it was realized the need of deciphering the system of classification, i.e. the syntaxes used in the language of the genes. In the mid fifties the Spaniard Severo Ochoa (1905–1993) had discovered, in the Biochemistry Laboratory of the Faculty of Medicine of New York University, an enzyme, called polynucleotide–phosphorilase, which allowed the artificial synthesis of small molecules of DNA which came to be decisive for unveiling the "genetic code." To that end certain ingenious experiments of controlled synthesis *in vitro*, using synthetic RNA of known composition made up with the Ochoa enzyme were carried out. This work was done by the Americans Robert W. Holley, Har Gobind Khorana and Marshall W. Nirenberg. The genetic code was unveiled when the nucleotide bases component of m-RNA were related with the aminoacids linked in the proteins produced *in vitro*.

The study of the code unveiled that the linear information in the genes, DNA segments, can be read without superposition and that the words in the genes language is composed of triplets of nucleotide bases. Every three successive bases constitute a "codon" and each codon has the meaning of an aminoacid, in such a way that, given that there exist 64 different combinations of 3 bases, i.e. 64 codons, and there are only 20 aminoacids, there are "synonymous" codons, i.e. codons which codify the same aminoacid. Of the 20 aminoacids, there are some which have at their service six codons, several which have four codons, and some two or one. Because of that it is said that the genetic code is degenerate. There are three codons (UAA, UAG, UGA) which do not codify for aminoacids, but are stop signals in the translation. They are codons which are located at the end of the sequence of one gene in such a way that, when the translation takes place, arriving to one of these codons the incorporation of aminoacids ceases and the protein chain is ready for processing to play the corresponding role.

The discovery of the genetic code is another of the important achievements of 20th century science and it astonished, for the fact of its universality, when it was checked that the same codification system works in the viruses, the bacteria, the fungi, the plants and the animals. This discovery unveiled other of the great secrets of nature, the unity in origin of all living beings: the monophyletic origin of life.

Almost in unison with the discovery of the genetic code, another fact was discovered, of great importance to understand the harmony in the ordering mechanisms which rules the expression of the genes. If one bacteria counts with more that 4,000 genes, each one capable of generating a protein, and if a more complex organism such as that of a mammal has above 20,000 genes, when, how, and why do genes express themselves?

It seems obvious that there should exist mechanisms that regulate its expression in such a way that each gene expresses itself only when and where the needed protein is lacking. For example, one bacteria living in a medium in which a type of sugar of whose existence the bacteria depends as a source of carbon, should be able to express the gene appropriate for the synthesis of the specific protein capable of metabolizing the sugar. In a multicellular an complex organism such as man, it is obvious that a cell belonging to a certain tissue would be synthesizing the protein or proteins characteristic of the specialty which that cell possesses.

In this way, if there are more that 200 cellular specialties, in each tissue will be expressing themselves only the genes proper of its specialty. So, a muscular cell will be expressing proteins proper of the muscle, one of the islets of Langerhaus in the Pancreas would express the gene of the insulin, in the fibroblasts would be synthesized collagens, etc.

Already in the sixties was assumed the existence of mechanisms of "regulation" for the expression of the genes. The discovery of how they work begun to be known thanks to the French geneticists François Jacob and Jacques Monod (1961), who unveiled the organization and the regulation of the expression of the genes in the bacteria.[4]

How do we explain evolution?

The fact that all living beings share the same coding system implies that the ancestor of all forms of life had already such a code at its disposal. From that "cenancestre"[5] would be issued the tremendous display of all life forms, through the phenomena of the increase of genes and of the genome size, through the accumulation of mutations and the diversification mediated by the need of adaptation to the mosaic of environments. Little by little, the different ecologic niches offered by our planet would be conquered. The capability of replication and mutation of the primitive DNA would allow the propagation and the enrichment of the various genes and genomes, the prime matter for the biodiversity, subject always to the refinement of its adaptive capabilities through natural selection, but keeping in any case the same coding system.

The Ucranian born geneticist Theodosius Dobzhansky (1900–1975), who was Professor of Genetics at the University of California and is considered the founder of Evolutive Experimental Genetics, said that: "nothing has sense in biology if is not in the light of evolution."[6] More recently, the Spanish evolutionists Francisco Ayala, disciple of Dobzhansky and Professor of Genetics of the University of California (Irvine) has pointed out that: "nothing has sense in evolution it is not in the light of genetics."[7]

Charles Darwin has pointed out that evolution is based upon "natural selection" which is the mechanism that explains the modification of the genetic characteristics of the populations along the generations. But in order to have selection is necessary first to have sufficient genetic diversity from which to select. That diversity is generated slowly but inexorably through the mutations. In this way the resources are created for the ulterior sieve of the best genotypes which are none other than those which best reproduce themselves, or, in order words those better adapted to the environment in which they live.

The theory of evolution as explained by Darwin lacked the understanding of two key phenomena needed to see how natural selection takes place, the "mutation" generator of diversity and the "hereditary transmission." The first step was described at the beginning of the 20th century by Hugo De Vries, who erred, however, when he gave too much importance to the mutations to which he proposed as the cause of the abrupt origin of new species in his "Theory of the Mutation." The second step was resolved with the rediscovery of Mendelism.

The interpretation of evolution in genetic terms enlightened what was later called "neo-Darwinism" which developed starting in 1930 thanks to the mathematical models put forward by Ronald Fisher[8], John Haldane[9] and Sewall Wright[10], which did show that the evolutionary capabilities of evolution of the populations is correlated with the existing amount of genetic variation in relation with adaptive capacity. Numerous experimental works and field observations have shown the veracity of these models. The evolutionary history of the living world is full of examples of extinct species as a consequence of the loss of relative genetic wealth relative to the environment because of natural causes or because human action challenges all the time the survival of the species, which results in the preferential reproduction of those individuals carrying a better combination of genes.

Today is admitted the definition of Theodosius Dobzhansky[11]: "The Evolution is a change in the genetic composition of the populations." As the Spanish evolutionist Francisco Ayala notes: "probably there is no other concept in any other scientific field which has been examined and corroborated more extensively and minutiously as the evolutive origin of the living beings."[19.]

How do we explain speciation?

An important question to clarify before explaining what genetics can say about speciation is that, contrary to a widely extended misunderstanding, speciation is not synonym of evolution. It may be evolution without speciation, while there will be no speciation without evolution.

During the evolutive process of a population or a species mutations are being produced which, if they are favourable in the environment in which they live, or are at least not unfavourable in the sense of damaging the biological efficiency and the reproductive capabilities of the individuals which carry them, become stored in the population. This means that, with the pass of time and generations, the individuals of the population will be genetically more differentiated from those of other populations of the same species submitted to different exigencies of adaptation. In other words, they will show in varying measure a set of heritable characteristics with respect to the starting generation. But the genetic distance is not enough for the two populations of the same species to become different species. When the flux of genes among groups of individuals pertaining to the same species is interrupted is when we can speak properly of speciation.

The main principles of this process and the precision of the meaning of speciation are due to a series of authors who put the bases of the so called "Synthetic theory of Evolution". In the first place it is necessary to cite Theodosius Dobzhansky, who, in 1937 published his work: *Genetics and the Origin of the Species*, in which he presented an integrated reasoning on the evolutive process in the light of the knowledge accumulated till that time.[12] It was also important the contribution of the German ornithologist Ernst Mayr (1905–2005), a man who reached the age of 100 years and which published in 1942 an essay entitled "Systematics and the Origin of the Species", which is probably the last classic contribution from the world of the specialized naturalists to the understanding of the evolution in relation with the development of genetics.[21] Two years later the American palaeontologist George G. Simpson (1902–1984) published another essay entitled "Time and Mode in Evolution" key in undestanding the union among palaeontology and population genetics.[13]

All these authors coincide in the assertion that a species is constituted by the set of individuals who share common genetic patrimony and they share it only with the individuals of the same group. When a species scatters its genetic characteristics transform till the apparition of two or more species speciation has taken place, which resorts in new taxons (cladogenesis). But it is also possible a gradual transformation of a species through time without diversification.

In this case speciation will take place by means of changes in the same lineage, which will be visualized in the morphology through time (anagenesis) becoming difficult to determine when the old species has become a new species. For instance, modern *Homo sapiens* is assumed to be derived from the evolution of other species of the genus *Homo* which existed before. In this case one would say that there has been evolution and speciation, which is more common, the difficulty lies in signalling the precise moment in which one species gives rise to other. This difficulty appears when there is no possibility of knowing if the individual before or after that given moment was capable or not of exchanging genes through reproduction simply because of they were not contemporaneous. One question which is immediately brought up is how the mechanism of interruption of genetic interchange among members of the same species takes place to give rise to two or more. To answer this question two models of speciation have been proposed, some of whose pointlike aspects have been corroborated in the laboratory or in the experimental field. It is important to note that the appearance of a new species is a rare phenomenon and it does occur under very special circumstances, in which various factors must happen. One of the factors, which must have contributed to the enrichment of biodiversity, is of geographic nature. The separation of a group more or less important of individuals of one species can give rise to a divergent evolution of the sets of individuals geographically distanced. Spatial insulation is already a forced interruption of genetic exchanges which, together with the phenomenon of cumulative mutations, genetic drift and natural selection under different conditions of adaptation, can result in an increasing genetic diversification between the separated sets of individuals. This geographic model of speciation was proposed by Ernst Mayv in 1963 for animal species.[14] In general, it is assumed that geographic insulation determines a progressive genetic differentiation of the groups, which, in the case of small populations, can be relatively rapid.

Once a process of speciation is underway a series of mechanisms of reproductive insulation, progressively more efficient are implanted to avoid the biologic waste involved in the genetic exchange between individuals in the process of separation. In species with sexual reproduction, the first manifestation of the process of separation consists in the loss of reproductive capacity in the crossbreeding between individuals that are becoming separated. To this end mechanisms of reproductive insulation become implanted which can be of two types: postzygotic and prezygotic.

In postzygotic mechanisms, there are still attempts of crossing between males and females of the groups in process of speciation. However, if there is already a certain degree of genetic distance, they will begin to emerge phenomena of inviablility, sterility, or degeneration of the hybrids.

The prezygotic mechanisms of reproductive insulation are more efficient because the crossbreeding is impeded and therefore is the formation of zygotes of hybrids. There are several types depending of the factor which enters in the impediment of the hybridization. Factors of temporary or stationary character, ethologic, mechanic or gametic can be mentioned depending on the impediments such as no coincidence in the epoch of sexual maturity, season, absence of sexual attraction between males and females of the groups under speciation, structural differences in the reproductive organs or lack of ability for the union of male and female gametes, respectively.

What is most interesting of all these mechanisms is that all these factors have a genetic basis, There are genes implicated in them, with allelic variations and differences in the phenotypic manifestations. As a consequence, the natural selection will improve the insulation mechanisms to hinder biological waste and to reduce the presence of forms with diminished or null fitness. Due to this, it is accepted that speciation begins when the postzygotic mechanisms show up and then will be substituted by other mechanisms more efficient, the prezygotic ones.

After the mechanisms of reproductive insulation between two groups of population there is no going back and the mechanisms of the incipient species would not return to exchange their genes resulting in increasing divergences as successive generations come forward.

Arriving at this point the question of the number of species which have existed in our planet from the beginning of life, some 3800 million years ago, could be rised. Life's history in our planet is the history of the appearance and the disappearance of species. It is assumed today that more than 99% of the species which were appearing along all that time have disappeared. About the species which are presumed to exist today, the number varies between two and five millions, according to different authors. The difference is too large due to the difficulties in knowing and mastering the totality of species presently alive. About animals, in the seas there are approximately 150.000 species alive, which is probably the highest number of species alive in all times. Animal life on land emerged about 400 million years ago in the period known as Devonic and probably is now in the maximum number of species. Of all the taxonomic groups, insects are the most abundant in the present moment, making up three fourths of the animal fauna and almost one half of the living species, including plants. Given that the insects appeared in the Carboniferous, 350 million years ago, we can conclude that in total, the insects are the animals with the greatest evolutive dynamics among all living beings. It seems certain that the phenomenon of life has expanded continuously and each time more accelerated towards the acquisition of different types and forms.

How is explained the diversity of design in the living beings?

Assuming the role of genetic variation, generated by mutations and natural selection, as the purifying mechanism determinant of the evolution of species, a question which found certain difficulties after the theory of evolution was accepted was the explanation of the sudden appearance of many very different species as revealed by the fossil record. The fossil series present modifications in the design of the body architecture, sometimes in periods exceedingly short. For example, why the placentary mammals have arrived in this day to more than 4,600 species pertaining to some 20 orders, which like "raying" different morphologic patterns, in 75 million years of evolution, while among amphibians, the frogs, in twice that time have resulted only in 3,050 species, pertaining to a single morphologic pattern?, why the production of types or biologic architectures has been so diverse in different evolutive stages and taxonomic groups? This phenomenon received the denomination of "macrooevolution." Today the term "macroevolution" is used to designate any evolutive change at or above the level of species, as the appearance of genus, families or higher taxa.

Jay Gould[15] and Niles Eldredge[16] tried to understand this phenomenon through the so called theory of "punctuated equilibrium," which sought to explain the drastic changes in biologic type without recurring to the intermediate "links" in the fossil record. These authors believe that macroevolution is due to cyclic processes of evolutive explosions with great changes, followed by periods with little changes during millions of years. These authors hold that the cladogenesis and the anagenesis are related causally and they postulate that, when a small population remains insulated, the species take more or less time to appear depending on whether they are a consequence of episodes of slow gradual change through natural selection in large populations, or whether they appear after the insulation of small groups in which genetic drift is acting. That is however, not a satisfactory explanation. It is required something other that the suppression or fixation of some alleles of the genes to explain drastic changes in the design of the organisms. Very different species cannot spring out purely random depending of the destiny of a group of more or less important genes related with the body structures.

The Genetics of Development explains the genetic mechanisms which enter in the morphologic transformations which are observed along the development of the living beings. This young branch of Genetics offers a satisfactory explanation regarding the decisions of organization of the body architecture of the multicellular animals and it can explain the phenomenon of macroevolution. It is necessary first to distinguish to that end between two types of genes, the "structural genes," directly responsible of the morphologic structure and from which depends the synthesis of the proteins determinant of the specific function of each cell, and the "regulation genes," which are those directing the expression of the structural genes in space and in time. The regulatory genes may be few in the whole of the genome, but they are very important, being those in charge of ordering which structural gene or genes, when and at which site of the organism must be active or inactive. The regulatory genes are those which dictate the orders while the structural genes are the ones which execute them, and depending on its activity, determine the cellular differentiation and the speciality of the tissue. The regulatory genes have the conductor´s "baton", in such a way that under certain stimuli dictate the necessary order for the activation of the necessary structural genes, only when and at the cells which are needed proteins they code.

In the last years it has been shown that the regulatory genes and the regulation mechanisms are practically the same in the whole evolutive scale of the metazoans. In this way, it has been found a surprising similitude in the structure and function of the so called "homeotic" genes, which constitute a type of regulatory gene implicated in decisions about the morphologic destiny of the cellular lineages and are extraordinarily similar in beings as distant as they fly *Drosophila melanogaster* and the mice.

All this means that a mutation altering the program would have a greater repercussion in the morphogenesis than a mutation in a simple structural gene. The mutations of regulatory genes that would advance, delay or impress an order in the genetic program of which depends the organization of an organ or a tissue, could give rise to a new design which, if it overcomes the filter of natural selection, would lead to a new type of organism. This would be the explanation why such diverse forms of life have arisen along evolution, keeping the same body elements and extraordinary homology in the organization of certain organs, as can be seen in the forward extremities of the vertebrates, or in the variations in the body organization of molluscs, or in the delay or advance of characters of one part of the ontogeny to other as in the case with the conservation of juvenile traits in the human adult with respect to those of the remaining hominids.

It could be said that the evolution is the result of changes in the genome along generations, being the evolution of the regulatory genome responsible of the macroevolutive phenomena and the mutations in the structural genome those which would determine slower gradual changes characteristic of microevolution.

How is the development of multicellular beings explained?

According to what was said in the previous section, morphogenesis is explained as a consequence of a program of genetic activities perfectly established in the zygote from the moment of the fecundation and it depends on the successive waves of expression of the structural genes under the command of the regulatory genes. The advances of Development Genetics and the capability of analysing the genomes of many species has revealed the relationship between the role of certain genes and the morphologic stages which are manifesting themselves phenotypically along the development of the multicellular beings. The genes which take part in the slow development of an embryo, the formation of gradients and morphogenetic fields determinants of the cellular differentiation and the organogenesis, are well studied in simple organisms such as the fruit fly *Drosophila melanogaster* or the nematode worm *Caenorhabditis elegans* and constitute one of the brightest pages of modern Biology.

Beginning with the fusion of the pronuclei of the feminine and masculine gametes begins the embrionary development through the execution of a program in which the regulatory genes are stamping each stage of development and determining the differentiation of the cells. In mammals, the whole process of development from fecundation to nesting and from gastrulation forwards is dynamic and takes place without solution of continuity with the differential intervention of certain structural genes present from conception but silent up to the moment in the development and in the embryo's place in which it is appropriate for them to express themselves to trigger the appearance of a type of structure.

In this way, for instance in the human being, the development is a process which goes through a series of stages from the zygote, going through a ball of a few cells, the morulle, the blastocyst that nests in the uterus (the 6th to the 9th day after fecundation), the gastrula and the phoetal stage. The successive phases do not represent any qualitative change in the developing embryo, but morphologic and quantitative steps in what refers to the size and the organization of parts.

At the beginning all cells are identical and "totipotent" which explains the fact of the ability to compensate the loss of cells or a subdivision of the embryo until the stage of blastocyst. In this way, an accidental subdivision in an incipient blastocyst explains the formulation of the monozygotic twins through which the parts of the original embryo which becomes independent could reorganize themselves and undertake the way of a normal and independent development. Towards the 4th day after the fecundation is completed the formation of three cellular strata: ectoderm, mesoderm and endoderm and morphogenesis begin.

After implantation in the uterus, as development advances, the totipotentiality is lost and the cellular lineages acquire their different roles through a process of cellular differentiation which implies activation or deactivation of different genes in each cell, tissue or organ, under the baton conductor of the regulatory genes. After fifteen days the general body plan is defined and organogenesis and hystogenesis begins.

In the same way as in the explanation of the macroevolution, the regulatory genes acquire an extraordinarily important role in the morphogenesis. It is evident that from the point of view of the hierarchy of the genetic activities which intervene during the ontogeny, the regulatory genes are more transcendent than the structural genes.

How are genomes organized?

The last quarter of the 20th century and the part elapsed of the 21th are marked by the advances of Molecular Biology, Computational Science and Biophysics, which have provided the necessary tools to approach the so called genome projects.

The genome can be defined as the whole of the genetic information contained in the molecules of nucleic acids of a living being (normally DNA, in certain viruses RNA). In the case of superior organisms the genome would be the global totality of the genetic information which exists in the nucleus of the initial cell, the zygote, formed after the fecundation, and it will be invariably conserved in each cell of an individual.

In the last years it has come up a new specialty, the "genomics," which is the branch of Genetics in charge of studying the organization and of analyzing the complete information of the genomes. The fundamental steps for the birth of Genomics were given after the discovery of the double helix structure of DNA and after knowing that the genetic information is stored in the succession of the nucleotide bases. Beginning in the second half of the decade of the sixties, researchers directed their attention to the organization of information in the genes, for which a series of techniques were developed to know the complete sequences of the genomes of some viruses with a very small genome. At the end of the sixties had been sequenced the genomes of the virus MS2[17] and ΦX174[18] which are the bacteriofagues which attack the bacteria *Eschechia coli* and whose genome is a simple molecule of RNA (3,569 bases and 4 genes) and a simple chain of DNA (5,375 bases and 9 genes, several of them overlapped) respectively. Soon afterwards it was completely sequenced the first genome of DNA of double helix, the SV40, capable of transforming human cells in cultivation and of inducing tumours in animals of experimentation.[19]

The idea of approaching the Project Human Genome began its gestation in 1984, at a Scientific meeting in Alta (Utah, USA) when an important group of researches discussed about the convenience of carrying out an ambitious program of high economic cost which would allow to know the genetic mutations causing the human sicknesses. Two years afterwards, in the Californian City of Santa Fe, the first proposal for a Project Human Genome was made as indispensable requisite to understand cancer. This was collected in convincing form by the researcher Renato Dulbecco in an important article in the journal *Science*.[20]

To approach the project it was necessary to develop a series of techniques capable of cutting in manipulable fragments the genomes for their conservation, clonation and posterior sequenciation. The cutting of the DNA is achieved with some special enzymes, the restriction endonucleares, discovered in the sixties by Werner Aber, Daniel Nathans and Hamilton Smith. The conservation and clonation of the fragments constitutes one of the most brilliant pages of genomics and it became possible through the invention of vectors of clonation adequate to maintain and replicate the fragments in the interior of micro organisms such as bacteriae and "yeasts". In this way a genome as large as the human genome, with its 3,175 millions of base pairs was maintained complete in small fragments in a collection of cultivations of thousand stumps of bacteriae or yeasts, each being the carrier of a little piece of genome which, as a whole, would constitute what has been called a "genomic library."

The sequence of little pieces of DNA (at least 700 base pairs) and their further ordering were important steps which required the perfecting of the sequenciation techniques in support of the bioinformatics. The molecular techniques which allow to know the organization of the genetic information and the sequences of nucleotide bases of the genes and the intergenic regions, enhanced the level of detail to which our knowledge of the genes was able to reach and opened the possibility of reading the complete genetic language of our species from end to end of each chromosome.

At the end of 1999 the journal *Nature* published the complete sequence of human chromosome 22,[21] the smallest of our genome, which contains 33 millions of base pairs, and the 8th of May of 2000, the same journal published the sequence of chromosome 21,[22] of great relevance for its implications in the Down Syndrome and other human sicknesses. A few months later, in February 2001, *Nature*[23] and *Science*[24] devoted special issues to publish respective articles on the results of the analysis of our genome and the main characteristics of its organization. Finally, on April 2003[25,26] in coincidence with the celebration of the 50th anniversary of the discovery of the double helix, Francis Collins, director of the public consortium of Project Human Genome, announced its culmination. Since then, till today, specialized works on the sequenciation of new genomes are appearing without interruption with special emphasis on the functional analysis of the genes and on the comparison of homologue regions of different species. The success of the Human Genome Project inspired other efforts of what came to be called "big-biology," mainly the Hap-Map Project, which allows us to abound in the differences between the different human genomes, and the encyclopaedia of the DNA Elements (ENCODE), which allows to identify each functional elements in the genome.

With all this information the geneticists have discovered that basic concepts such as "gene" and "genetic regulation" must be considered today in a more complex fashion than it was assumed at the beginning. Today we have learned that in all events of the plan for the development of the body of a higher organism, as a mammal, underlies a complex system of genomic regulation, and that the changes in this system utilize the same type of genes in the different living species, which corroborates what Development Genetics had anticipated as an explanation of "macroevolution." Also, trough the Genome Projects, the possibility has been opened of understanding how a living being is constructed, starting with its genome and of "disentangling" the evolutive relationships between different species. These processes have propitiated also the opening towards a series of biotechnological applications of great social and economic interest.

We can ask whether all this intellectual ferment has benefited really human health. Probably the most important think about the Project Human Genome is yet to come, in connection with the biomedical applications in three fields: the diagnostic, the therapeutic and the pharmacological. Notable in this connection are the advances in the field of drogues directed against identified genetic defects, in a few types of cancer and in some rare genetic sicknesses. The comparison of sequences of given regions of the genome of patients affected by a certain sickness and healthy persons is the foundation in which rests the diagnosis of many pathologies and the testing ground to approach a personalized medicine.

Francis Collins in an article in *Nature* in April 2010 notes that 10 years after the acquisition of the first "draft" of the human genome: "the success of personalized medicine will depend in the exact identification of the risk factors, genetic and environmental, and in the capacity to use this information in the real world to influence the health behaviour, and to achieve better results. This requires new research projects in grand scale well designed to discover the risk factors and to probe the implementation of prevention and pharmacogenomics programs."[27]

Integrating role of Genetics in Modern Biology

The advance in knowledge has been extraordinarily ordered and rapid. In little more than one hundred years we have come from not knowing anything about the seat of genetic information and its molecular nature, to know the complete genetic information of many species.

We can conclude from this trip that the central role of Genetics in Biology has been made possible thanks to the knowledge of the DNA, defined as the "molecule of life." This molecule exists in fact in all living beings and it fulfils a role in the transmission of information from cell to cell, from parents to sons, from one population to another and from one species to another along time. It constitutes the conductive thread of life of each organism and of the whole of organisms.

The DNA is in the centre of all explanations of the biologic phenomenon. The Italian geneticist Salvatore Luria (1912–1991), Nobel Prize in Medicine, 1969, said that "life differs from all other natural phenomena in a fundamental trait: it has a program."[28] As the program implies a series of ordered operations to carry forward the project, it is obvious that the map of the program resides in the DNA. In the same way, the British evolutionist John Maynard-Smith (1920–2004) centered the difference between living and non-living beings in the properties of DNA when he noted that: "The entities are alive if they have the properties necessary to evolve by Natural Selection. In other words, if they can multiply, if they possess a mechanism to insure the continuity of character and if they can vary."[29] Also the cellular biologist Nobel Prize of Medicine 1974, Christian Duve, says that "life is what is in common to all living beings," and that: "The information which guides the ensambling of the proteins is not given by the proteins, but by the nucleic acids. And these are the molecules which place themselves at the top of the hierarchy of the organization of the cells."[30]

Perhaps the privileged position of Genetics among the science of our days is due to have explained better than any other branch of Biology the inherent characteristics of the living beings. However with all which is known till now, we would be explaining only the properties of the living beings, but not the causes of the phenomenon of life, neither the meaning of life. It would remain the question: What is life? A question which is not proper of science, and whose answer enters in the realms of the metabiology, the metaphysics and even the theology, but, in any event, is left out of the space of the scientific method.

References

1. Ch. Darwin. *The variation of animals and plants under domestication.* London: John Murray (1868). Retrieved 1 November 2008.

2. Ch. Darwin, *On the Origin of Species by Means of Natural Selection,* or the *Preservation of Favoured Races in the Struggle for Life,* John Murray, London 1859.

3. J.D. Watson, F.H. Crick, "Genetic Implications of the Structure of Deoxynucleic Acid", *Nature,* 171 (1953), pp. 737–38.

4. F. Jacob, J. Monod. "On the regulation of gene activity". Cold Spring Harbor Symp. *Quant. Biol.* 26: 193–211.

5. Last Universal Common Ancestor.

6. Th. Dobzhansky, "Nothing in biology makes sense except in the light of evolution", in *The American Biology Teacher* 35 (1973), pp. 125–129.

7. F.J. Ayala. *La evolución de un evolucionista.* Col. Honoris Causa, Univ. de Valencia 2006.

8. R.A. Fisher, *The Genetical Theory of Natural Selection,* Clarendon Press, Oxford 1970.

9. J.B.S. Haldane, *The Causes of Evolution,* Harper, New York 1952.

10. S. Wright, "Evolution in Mendelian populations", in *Genetics,* 16 (1951), pp. 97–159.

11. Th. Dobzhansky, *Genetics and the origin of species.* Columbia, New York 1951.

12. E. Mayr, *Systematics and the Origin of Species.* Columbia University Press. New York 1942.

13. G.G. Simpson. *Tempo and mode in evolution.* Columbia University Press, New York 1944.

14. E. Mayr, *Animal Species and Evolution.* Harvard University Press 1963.

15. S.J. Gould, Wonderful Life: The Burgess Shale and the Nature of History, Norton 1989.

16. N. Eldredge, Macroevolutionary Dynamics: Species, Niches and Adaptive Peaks, McGraw-Hill, New York 1992.

17. W. Fiers W, R. Contreras, F. Duerinck, G. Haegeman, D. Iserentant, J. Merregaert, W. Min Jou, F. Molemans, A. Raeymaekers, A. Van den Berghe, G. Volckaert, M. Ysebaert. "Complete nucleotide sequence of bacteriophage MS2 RNA: primary and secondary structure of the replicase gene". *Nature* 260 (1976) 5551: 500–507.

18. F. Sanger F, GM. Air, BG. Barrell, NL. Brown, AR. Coulson, CA. Fiddes, CA. Hutchison, PM. Slocombe, M. Smith." Nucleotide sequence of bacteriophage phi X174 DNA". *Nature* 265 (1977) 5596: 687–695.

19. W. Fiers, R. Contreras, G. Haegeman, R. Rogiers, A. Van De Voorde, H. Van Heuverswyn, J. Van Herreweghe, G. Volckaert & M. Ysebaert "Complete nucleotide sequence of SV40 DNA" *Nature* 273 (11 May 1978), 113–120.

20. R. Dulbecco, "A turning point in cancer research: sequencing the human genome", in *Science*, 231 (1986), p. 1055.

21. I. Dunham,.y col., "The DNA sequence of human chromosome 22", en *Nature* 402 (1999), pp. 489–495.

22. M. Hattori, and col. "Chromosome 21 mapping and sequencing consortium. The DNA sequence of human chromosome 21", *Nature* 405 (2000), pp. 311–319.

23. The International Human Genome Mapping Consortium "A physical map of the human genome", in *Nature* 409 (2001), pp. 934–941.

24. The Celera Genomics Sequencing Team. "The sequence of the human genome", in Science (2001), pp. 1304–1351.

25. F. Collins, E. Green, A Guttmacher, M. Guyer, "A Vision for the Future of Genomics Research. A blueprint for the genomic era", in Nature 422 (2003), pp. 835–847.

26. F. Collins, M. Morgan, A. Patrinos, "The Human Genome Project: Lessons from Large-Scale Biology", in Science 300 (2003), pp. 286–290.

27. Francis Collins, "Has the revolution arrived?" *Nature* 464, 674–675 (1 April 2010).

28. S.E. Luria, *La vida, experimento inacabado.* Alianza Editorial, Madrid 1975.
29. J. Maynard Smith. "Time in the evolutionary process", en *Studium Generale* 23 (1970), pp. 266–272.
30. Ch. De Duve. *La vida en evolución.* Crítica, Barcelona, 2004, p. 25.

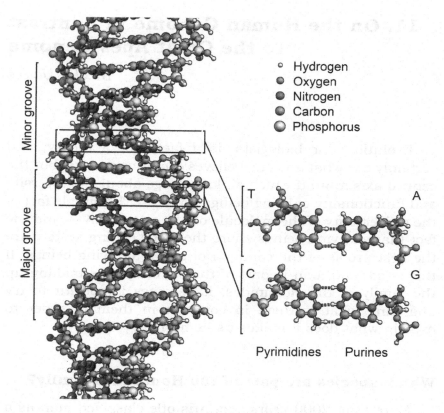

Hydrogen
Oxygen
Nitrogen
Carbon
Phosphorus

Minor groove

Major groove

T   A

C   G

Pyrimidines   Purines

The structure of the DNA double helix

11. On the Human Genome in Contrast to the Great Apes Genome

Nicolás Jouve

Evolution, for biologists, is a fact, not a theory, and certainly not what any one believes evolution means, but the central axis around which all knowledge about the diversity and functionality of living beings revolves. The simple fact of the universality of the molecule of life — the DNA — and the fact that all living beings share the same coding system are the best proofs of the common origin of all living beings. If this is so, and man is part of the set of species making up the family of the Hominides, it becomes important to try analyzing what is there in common in them in order to explain what does it makes us be human.

Which species are part of the Hominides family?

More than 2000 years ago, Aristotle classified man as a hot blood animal, noted the enormous proportion of his brain, and underlined his intelligence and his capability of relating with his fellow men saying: *"man is a political animal"*. Anthropocentrism was a constant since that time among all successive cultures, underlying man's origin as something planned by the Creator from the beginning, as a created in his image and likeness. In the 18th century, Swedish naturalist Carl Von Linnaeus (1707–1778) published his great book entitled *"Sistema Naturae"*, an attempt to order and classify all living beings for which he proposed a hierarchical method of classification for all species of plants and animals, based upon the morphological characteristics or other observable properties. Linnaeus made a single slot for the human species for which he proposed the generic denomination *Homo* and included in it a single species, the *Homo sapiens*, which did show that Linnaeus ascertained the ability to ratiocinate as the distinctive element of the human being. He proposed also the grouping of three families in a superfamily, called Hominoidea (hominids) which included besides man the Pongidae (pongids) which grouped those apes closest to him: chimpanzees, bonobos, gorillas and orangutans, and the Hylobatidae (hilobatids) which include the gibbons and Asiatic Simians. Linneaus grouped all these species in the order Primates (Apes), meaning the first in nature. The grouping responded to merely descriptive aspects, but, after the adoption of the evolutionary theory two centuries later, the grouping in a single superfamily is reinforced by the fact that these families constitute a phylogenetic group which descends from a common ancestor.

Already in the 19th century, Charles Darwin included formally the human species in the same evolutionary scheme of all species, first in 1859, in his principal work *On the Origin of the* Species by Natural Selection[1] and later in his book *The Descent of Man and Selection, in Relation to Sex*[2] in that he proposed the evolution of the man from similar ancestors to monkeys. Nevertheless, it is necessary to say that this does not mean that "the man descends from the monkey", as sometimes the idea of Darwin has been interpreted. What Darwin exposed with total foundation is that the present man and big apes had a common ancestor in the past.

Modern taxonomy, based upon evolutionary diversification, groups within the Hominidae the great Simians previously classified as pongids (chimpanzee, gorilla and orangutan). Most present scientific work classifies the biped Hominids as making up the subtribu Hominina. In what follows we will use the generic terminology hominids for the grouping of man and the primitive pongids. Today it is generally accepted, in any event, that this group of species follows from a common ancestor and that they are the result of the diversification in divergent evolutionary lines. Among them, the nearest to the human beings is that of the chimpanzee, *Pan troglodytes*, and his near relative the bonobo, *Pan paniscus*, from which line it departed more than 6 million years ago. In turn, humans and chimpanzees departed from the line leading to the present Gorilla, *Gorilla gorilla*, about 10 million years ago, and the group of all these species may have diverged from the orangutan, *Pongo pygmaeus*, about 15 million years ago.

Once divergent, the evolutionary line leading to modern man presents a series of links, known in part, after the discovery of a series of remnants of primitive forms which show a gradual evolution in a series of morphologic traits early identifiable away from the monkeys and approaching *Homo sapiens*. It is relevant to note that fossils constitute irrefutable proofs of the gradual modifications of those organisms along time. It will be possible to discuss on the concrete validity of certain fossils, or on the exact age that they have, or its temporary order within the same evolutionary line, but they constitute unquestionable tests of the evolutionary process.

During the process of human biological evolution — hominization — it is possible to study in the remnants of pre–human fossils the alimentary habits, dentistry, walking manners, erect disposition, encephalic volume and other physical traits. The direct ancestor of the genus *Homo* appeared in a geological period characterized by the cooling of their African habitat in the Pliocene and the middle Pleistocene, about 5 million years ago. Excavations carried out in numerous sites at South and East Africa have lead to the discovery of a series of primitive forms precursor of *Homo sapiens*. If we limit ourselves to the last 3 million years, important links are the *Australopithecus africanus*, from which fossil remnants have been found in South Africa, with an antiquity of between 2 and 3 million years, the *Homo habilis* (between 2.3 and 1.6 million years), with diverse findings in Kenya and Tanzania, and *Homo ergaster* (between 1.9 and 1.0 m.y.), whose remnants have appeared also in Eastern Africa, which is assumed to be the antecessor of another important species, the *Homo erectus*, which lived from about 1.6 million years to about 50,000 years ago, and would be the first species of the genus *Homo* who went abroad from the African continent, even though is not considered a direct ancestor of *Homo sapiens*, but a collateral line.

Studies presently being carried out on predecessors of human beings in an epoque as close as between 900,000 and 200,000 years in the side of rocky mountain of Atapuerca near Burgos (Spain) are noteworthy. This site is very rich in remnants of great importance. Among the fossils encountered in the "abyss of the bones" and in the "great doline" are located the remnants of the oldest ancestors of man in Europe, the *Homo antecessor*, considered as a species which lived in the early Pleistocene (between 850,000 and 750,000 years of antiquity) which went through Europe from Africa and would be the last ancestor common to *Homo sapiens sapiens* and *Homo sapiens Neanderthaliensis*. This last species would have evolved from the *Homo Heidelbergiensis*, an intermediate type who would have also been found in Atapuerca. Among the best preserved fossils of *Homo antecessor* is a jaw which must have belonged to a child of about 10 years found in the "abyss of the bones" dated between about 780,000 and about 857,000 ago.

Without trying to do an exhaustive evaluation of human evolution, the findings in various sites demonstrate that 50,000 years ago three species of *Homo* existed on the surface of the Earth. They were extended all over and even coexisted on different regions. In Europe, the Near East and central Asia lived the *Homo sapiens Neanderthaliensis*, in the Far East, in particular in the island of Java, remained the last representatives of *Homo erectus*, and in Europe, coexisting with the Neanderthals, was the *Homo sapiens sapiens*. The three species had come from Africa and had passed to Europe and Asia through the Near East. With the extinction of the Neanderthals, about 30,000 years ago, *Homo sapiens sapiens* remained as the only representative of the genus *Homo*. The amazing qualities of this species allowed it to conquer all continents and habitable sites on our planet.

The evolutive history of the species is reflected as much in the fossil remnants of their ancestors as in the geographic distribution of the present species. In this way, the orangutan, which is the species more distant philogenetically of the presently current hominids, is also the one which lives in the area more distant. Its habitat is in the humid jungles of Indonesia and Malaysia, where human influence, regrettably, has reduced the number to about 35,000 individuals living free, mostly in the island of Borneo. From the remaining species, the *Gorilla gorilla* and the chimpanzees live in the tropical humid jungles of the African plains and mountainous regions. Of them, the small bonobo only exists in the Congo and its situation is of great risk of extinction, with a free population of less than 15,000 individuals in their natural habitat at present. The oldest fossil records of the genus *Homo* have been found also in the African continent, with important findings dispersed throughout various sites of South and East Africa.

How to recognize the special value of human life

When we see the changes operated in the evolution of the species of hominids we observe that they have been significantly more important in the human species than in any other species of the so called great apes, and more so than in any other species in nature. The evolution of brain volume and the ability to reason were in parallel with other important biological innovations, such as the improvement in morphologic gracefulness, bipedism, liberation of the hands, reduction of the teeth, and the acquisition of symbolic and articulated language.

Something which is evident among the species of the animal kingdom is that the human species is unique in the development of the intellect and in being the only species in which each individual is conscious of its own existence, a characteristic which we define as self-consciousness. To this is added a second unique characteristic in the context of nature, the capability to communicate by means of the language. This capability does not consist in a simple oral interchange based upon the possibility of speaking and hearing, but upon that of transmitting and interchanging ideas. This is what we mean by language of double articulation. It consists of a communication by means of words understood as sounds, in such a way that the words have a double function: they are sounds with meaning and they are ideas converted into sounds.

Speech and artistic creation should be considered the result of a parallel anatomic evolution needed to approach both functions, the throat's supralarigean region and the intelligence have their seat in the working of the brain. Since artistic creation is apparent only in modern man, *Homo sapiens sapiens*, in East Africa since about 100,000 years ago, and above all, in the art explosion in the Higher Palaeolithic in Europe, a little more than 30,000 years ago, we can locate in this period the acquisition of articulated language. According to this, neither the Neanderthals nor any other previous species of the genus *Homo* could have enjoyed the capacity of speech or artistic creation.

It should be taken into account also that human evolution is accompanied by a process of encephalization, a directional and accelerated evolution in the last 150.000 years towards the acquisition of a neocortex increasingly greater, which would contribute to the development of such important functions as sensory perception, the generation of motor commands, spatial reasoning, conscious thinking and language. In human language are implied not only the anatomic organs of throat and mouth but also the brain, specially the prefrontal and temporal regions, in particular the left temporal region, which is the one in which the nervous centre coordinating language is located. In this way our species acquires the singular capacity of vocalization, the cognitive faculty, the abstract reasoning and the self perception as an independent being. It is during this evolutive process of encephalization during which rational intelligence develops and is added cultural evolution to biological evolution. As Kieffer notes *"human beings do not live already in a physical world but in a symbolic world"*[3].

When we refer to the difference between man and the remaining animals it may be absurd to underline exclusively the physical traits insisting in man's animalism, as done with excessive vehemence by the British zoologist and etologist Desmond Morris, author of *The naked Ape*, in 1967[4]. It is ridiculous to consider the absence of hair as the best difference between man and the rest of primates. Perhaps it would be better to speak of "The Dressed Ape", underlining in this way man's peculiarity in covering his body with external elements in which psychological, cultural, artistic and protective traits are mixed, showing again the singularity of the human species.

In this way, man is able to perceive the entourage in which he lives and this conscientious way of life elevates human species to a higher dimension. The cultural evolution — humanization — is added to biological evolution. As a result man does construct utensils, learns to use fire, gets clothes for his protection, fabricates refuges, acquires dexterity for hunting and defence, tames wild plants and wild animals, organizes himself in groups, emigrates, controls other species, conquers all kinds of environments and adds new qualities to his presence in nature, the meaning of ethics and transcendence.

Every human being gathers experiences along his life, interchanges those experiences with those persons around him and transmits them to his offspring, passing them on in this way from generation to generation. According to F. Ayala,

Biological heritage is in man similar to that of other organisms endowed with sexual reproduction and is based upon the transmission the genetic information coded in the DNA by means of the sexual cell from parents to children. On the contrary, the cultural heritage is exclusive of humans and transmits information by means of a process of teaching and learning which is in principle independent of the biological inheritance[5].

It is possible that other species, of which some are of the opinion that they are intelligent, might be able to store experiences in their memory, but they lack the capacity of analysing them, or transmitting them, or acting about them with free will. At most they can store them for themselves, register them and react instinctively in future occasions.

What is the difference between man and the remaining hominids?

Besides everything said till this point and about the evolution of a material body with differences and similarities with respect to the other hominids, a new and distinct reality must be recognized in man: the pertenence to him of an immaterial spirit hypostatically united to the body. In each human being two realities coexist, body and soul, not a simple association of two different entities, but as a single entity. In this respect the document *Dignitas Personae*, sponsored by Pope Benedict XVI and published at the end of 2008 by the Congregation for the Doctrine of the Faith, says:

> "Although the presence of the spiritual soul cannot be observed experimentally, the conclusions of science regarding the human embryo give "a valuable indication for discerning by the use of reason a personal presence at the moment of the first appearance of a human life: how could a human individual not be a human person?"[6]

The twin corporal-spiritual dimension of man is reflected in his conscious way of facing life. The faculties of reasoning, self-consciousness, free will and self-restraint are its fruit, which translates into his capacity, unique among all living species, of facing life in a personal way. In order to evaluate human life it is very important to distinguish the privileged position of the human being in the context of creation.

The acquisition of a conscience by means of the incorporation of the spirit stirs up in man the needs to know and to get answers to questions which are not raised in any other species. About this singular characteristic of the human species, Carlo Rubbia, physics Nobel Prize winner, 1984, says:

> "the greatest manifestation of freedom is that of being able of asking himself where we come from and where are we going to...There is no class of human life which has not asked itself that question. And there is no human society which has not tried in some way to get an answer. To fail doing so is a loss, a becoming less human, a self-imposed punishment"[7].

All this implies a great difference with what takes place in the most advanced animal species. The meaning of the peculiarity of our species consists in that human life is raised to a very special higher dimension which connects it with a dignity and a specific superior value. According to this no human individual can consider himself as an anonymous member of a given biological species submitted to an inevitable life cycle, but as a being which lives fully consciously his existence and is the protagonist of his own biography, which is unfolding according to his personal decisions, fruit of his own will and of the surrounding circumstances. At variance with the individuals of the remaining species, man is one who decides and so builds up his own being, not one who simply exists.

How can we compare man and the remaining hominids?

For many years, researchers have tried to investigate the differences and similarities within the family of the hominids in order to understand the biological key which triggered a conscious, speaking and cooperative being, starting with a primitive beast driven solely by instinct and self-interest. According to F. Ayala:

"The case of humans is particularly interesting because their ability of perceiving the environment and reacting to it flexibly is one of the most fundamental differences which set apart human beings from the rest of the animals..."[8].

The truth is that the great differences between man and his nearest relatives are not physical or biological, but of a more subtle and extraordinary kind. Accordingly, it is important to analyse in advance what we wish and we can compare in man and in the remaining species of the family of the hominids. While in human beings we recognize two realities hypostatically united, one spiritual and one material, in the animals most alike to man we recognize only a corporal reality. Regardless of our wishes our comparisons must limit themselves to that which is common to both, the material reality, which is also the only experimental reality, susceptible of being analysed.

The question is therefore, whether the comparison of the elements common to man and to the remaining hominids can give an explanation of the most singular properties of man, those which make us human. In what follows we will see that it is possible, through the study of the genomes, to explain the reason of the human singularity in some of its most notable properties, such as the cerebral functionality or the ability to speak. It remains concealed, however, to the scientific method, the spiritual component, unassailable to us from the experimental viewpoint.

The human genome project

The Genome Project started in 1990 and had for its goal to unveil the plethora of information enclosed in the human genes, coded in the sequence of nucleotides of the DNA. The aim was also to analyze the variations in that information in order to explain the pathologies and the diversities among individuals and populations. In February 2001, the prestigious journals *Nature*[9] and *Science*[10], devoted special issues to publish respective articles on the results of the analysis of our genome and the main characteristics of its organization. In April 2003[11,12], in coincidence with the 50th anniversary of the discovery of the double helix, Francis Collins, principal investigator and director of the International Consortium organized for the study of the human genome, announced the culmination of the Project. An effort of 13 years of deciphering the coded information in the human chromosomes, sheared by tens of laboratories in half a dozen countries had concluded. In this way a new era was inaugurated in biology, the «genomics» era which would allow investigation of the main biological phenomena in a more direct and more objective way. No doubt one of the great applications of this new discipline would be to undertake the phylogenetic relationships between the species by means of what has been called "comparative genomics".

In fact, the Human Genome project has been the testing ground for the approach to the knowledge of the genomes of other species. After its culmination, it was demonstrated the power of knowing the genomic maps, the catalogues of genes and the DNA sequences of any species and, in passing, the possibility of making comparisons to explain differences and similarities and to study the philogenetic relationships. Already from the beginning of the Human Genome project work teams and consortia among laboratories of various countries were organized to study the genomes of other species pertaining to other taxonomic groups. During the last years the genome projects of a set of model species have been finished, going beyond the original previsions. In June 2012 the great data collecting center in which genomic studies are gathered (http://www.genomesonline.org) registers the completion of the genome projects of 3373 species, which represent all taxonomic groups. From them, 3197 species are prokaryotes (154 *Archaea* and 3043 bacteriae) and the 176 remaining species correspond to superior eukaryotes, among them several species of primates and all the species of hominids. It is interesting to note that besides the species living at present, fossil remnants have allowed the extraction of their DNA in good conditions, as is the case with the neanderthals to which we refer in the next section.

The Genome of the Neanderthals

A question of great interest formulated by the anthropologists since the discovery of the contemporaneity of modern man, the neanderthals and the *Homo erectus*, is that of whether they were true species, separated by effective barriers of reproductive isolation, or the proximity of their biological origin was so lax as to allow certain reproductive exchange and therefore genetic. The differences between *Homo erectus* and *Homo sapiens* are large enough to attach to both taxons the rank of separate species and therefore suppose that there was no genetic exchange between them, at least till studies of comparative genomics are made. However the studies of the genomes of modern humans and the neanderthals suggest the possibility that crossbreeding was produced between both species.

Some years ago the team headed by the Swedish geneticist Svante Pääbo, who carries out his activity in the Institute Max Planck of Evolutive Anthropology in Leipzig (Germany) undertook the study of the genome of the neanderthals from the DNA of fossil remnants of this species. The main goal of this work was to unveil the genomic relationships and therefore the evolutive proximity and the possible genetic exchange of neanderthals and sapiens.

In their work Pääbo's team used samples of remnants found in Europe and in the Siberian location of Danisova for further comparison with the genomic DNA of present man, choosing in this case DNA samples of human populations of Europe, Africa, Asia and Papua Guinea. The comparative study of the genomes of these forms seems to demonstrate that the Danisovan neanderthals were able to crossbreed with the sapiens and from them the imprint of 6% has been left in the present human populations of Melanesia and North and North East Australia, and up to a 4% of DNA in the genome of Western humans[13].

The professor of immunology Peter Parham[14], of the Faculty of Medicine of Stanford University, notes that the genetic exchange between neanderthals, Europeans or Danisovans, and the sapiens must have had a positive effect in the acquisition of immunologic defences by modern men, through the introduction of new variants of the genes of the system HLA. So the variant known as HLA-B 73, which was found in the genome of the Danisovans, is rare in the present populations of Africa, but it is found more frequently in the present human populations near Danisova. This suggests that probably this genetic variant proceeds from the crossbreeding of sapiens and neanderthals in that region in which they coexisted 30,000 years ago. Another type of gene, called HLA-A 11, is absent among the African populations, but it represents up to a 64% of the variants of the gene present in the current human populations of East Asia and Oceania. These genetic systems could have been introduced in these populations through eventual crossbreeding with the archaic neanderthals and could have been kept by natural selection in modern humans as they conferred to them advantages for a better defense against pathogenic agents specially viruses.

The absence of the alleles of the above mentioned HLA systems in the present African human populations would be due to the crossbreeding of the ancient sapiens with their neandertal relatives after their migration from Africa in the sites where they coexisted no less than 65,000 years ago.

Comparative Genomics among Hominids

The human genome and the genomes of the remaining hominids posses an amount of DNA very similar, and a little above 3.1 million base pairs (double helix steps). Due to its origin from a common ancestor they keep a high degree of homology in the orthologue regions of their genomes and a like number of genes, which in man is estimated around 21,000. The genome projects offer a unique opportunity for analyzing the differences in those genes or regions of the genome which are common to the species which one wishes to compare.

In 2005, a couple of years after the completion of the human genome project it was completed the project chimpanzee genome[15]. In January 2011, researchers at the Genome Center of Washington University announced the completion of the study of the orangutan genome[16], and at the beginning of 2012, a scientific team directed by Aylwyn Scaly and Richard Durbin, of the Welcome Trust Institute (United Kingdom) announced the culmination of the gorilla genome project. From this knowledge it is possible to compare the sequences among the DNA sequences of homologous regions of all species of hominids[17].

Already since the sequentiation of the chimpanzee genome was in progress, prestigious journals like *Nature, Science, Genome Research*, etc, begun to publish comparisons of the homologue sequences of these species with that of the human being including regions partially known of the gorilla genome. The initial results pointed to differences around 2% among the three species, while there was emphasis in the existence of genes and regions of the genome with more or less homology. In one of the first comparative studies, researchers Chen and Li had published that from a total of 115 millions of orthologue nucleotide base pairs distributed at random over the genome of the three species, the human sequences were 98,76% identical to those of the chimpanzee, and 98,38% identical to the sequences of the gorilla[18].

It is necessary to take into account that in 3.100 million base pairs a divergence of 2% translates into about 62 millions of point differences in the nucleotidic bases. Leaving aside the 58% of the genome of these species, whose DNA sequences are not implicated directly in the synthesis of proteins or in the regulation of the genetic expression, the remaining 42%, which would be the really implicated in the functional jobs, accumulates much variation. It is here where the explanation of the differences between man and our closest relatives should be looked for. The data provided by the genomic comparison show that there are many mutations of changes, losses or additions of bases in these regions, which can induce differences of amino acids and of functionality in the coded proteins.

After the sequentiation of the gorilla genome it is verified the high degree of coincidence of the sequences of the DNA of the genomes of this species with respect to those of chimpanzee and man. Also it is confirmed that the greatest part of the genomes of chimpanzee and man are very similar; there is at least a 15% of the human genome more alike to that of the gorilla than to that of the chimpanzee, while other 15% of the genomes of the two species of simians appear to be more coincident with each other than with the human DNA. This high degree of genomic similitude supports the idea that the appreciable differences in the qualities that make us humans should not be as much in the structure of our genes as in the details of apparently small differences in its functioning. In other words, two distinct species with the same genes, can show behaviours or biological qualities different if they diverge in the levels of expression, in the moment of the development in which the genes act and in the functional interaction with other genes.

Another aspect of the interest in the comparative study of the genomes of related species has to do with the rate or speed at which changes take place after the evolutive divergence takes place. It is what is denominated the «molecular clock», that can directly be quantified when quantifying the differences of nucleotides of the DNA or the amino acids of proteins which is expressed with a simple parameter known as «mutacional distance». This parameter reveals the number of modifications by unit time. This type of analysis offer very interesting results. Thus, the genes related to the auditory sensorial perception and the development of the brain of the three species show as a whole an accelerated evolution with respect to other regions of the genome, being particularly high in the human beings and gorillas.

Moreover, evidence exists on the accelerated rate of modifications in certain regions of the human genome with respect the homologous regions of chimpanzee, gorilla and orangutan. Between these regions have been implied genes and DNA sequences involved in functions as important as the capacity of oral communication, the intercellular transmission of nervous signals or icohesions. The importance of these changes is fundamental to explain the degree of evolutionary specialization at which each species has arrived and very particularly the spectacular brain development and human intelligence. Svante Pääbo has carried out a meticulous analysis of a series of regions of the genome. Thus, he observed a simple mutation in the gene *Neu5Gc*, in *Homo sapiens* called *Neu5Ac*, that has repercussion in the immune system, the protein composition of cellular membranes and the establishment of intercellular connections, with direct consequences in the brain functionality. In view of these results, Svante Pääbo indicated that *"the human brain has accelerated the use of the genes"*[19].

Similarly, the group of Katherine Pollard of the Glandstone Institute of the University of California, in San Francisco, has registered the existence of 49 regions of the human genome, known as "HAR", (Human Accelerated Regions), that show a specially high accumulation of mutations unlike what happens in the genome of the chimpanzee. According to Pollard these regions have been favored to an accelerated variation by natural selection, with positive functional consequences. Thus, in a short region of 118 nucleotides of "HAR-1", present in chromosome 20 human, there is an accumulation of up to 18 changes with respect to the homologous sequence of the chimpanzee genome. The singularity of this finding lies in that this region is implied in the neuronal activity, and plays a key role in the development of the brain crust[20]. The accelerated evolution of regions HAR of the human genome must have contributed also to a greater compaction of the neurons, with repercussions in the metabolic levels of oxygen consumption, so that a smaller consumption of energy favors the functional capacities of the human brain.

Moreover, the group of Svante Pääbo observed a rate of accelerated evolutionary change in the coding region of the human gene *FoxP2*[21]. This gene exists in all the vertebrates and is involved in the capacity of speech and other neuronal functions. *FoxP2* codifies for a regulating protein that, in the human case, displays the substitution of two amino acids, not modified in the homologous gene of the remaining hominids. This protein, along with other coded by other genes, is implied in the specific human ability of the articulated language. Being this capacity one of the great singularities of the human being, this is the best example of how a small change in a simple gene can confer properties of control of the environment and exploit the relation with the other members of the species. This has direct consequences in the human capacities of creativity, something substantial in the development of the abstract reasoning, intelligence and the transmission of experiences.

If in the DNA we are so similar... What makes us humans?

The reported investigations show how small changes in the genome can determine great phenotypic differences and new capabilities, which, in the human case, result in the appearance of self-consciousness and in a great increase of the capability of communication among individuals and generations. In other words it is no as important the large percent of DNA in common as the small percent of difference in DNA.

At this point a reflection is in order. The knowledge of the genome allows us to register the biologic substrate implied in the evolutive changes, the adaptations and the capabilities of some beings with respect to others, but, is this sufficient to explain the human singularity? The family of the hominids has generated numerous species from which only *Homo sapiens* shows the superior capabilities and the singularity of which we spoke above. All hominids living today, the so called great apes and man, share a common ancestor who must have existed more than 15 million years ago and which came to be the necessary substrate for the appearance of an intelligent, reflexive and ethical species among a series of brute species. Could chance and natural selection be enough for this? And why only one species has reached the characteristics which define humanity? Biological evolution has been equally possible for all species, with the same mechanisms, as the fossil record and the genomic analysis shows. But a being as distinct and special as *Homo sapiens* seems to require an explanation which transcends scientific knowledge. The evolution of the great simians [apes], previously called pongids, did never reach the inflection point necessary to conquer the reflective capabilities which give rise to cultural evolution. Their evolutive ceiling remained at a level which is very far from that of self-conscience, and ethical behavior arrived at in human evolution. The biological substrate of the hominids could be the clay which served the Creator to produce a being made in his image and likeness.

References

1. Ch. Darwin, *On the Origin of Species by Means of Natural Selection*, or the *Preservation of Favoured Races in the Struggle for Life*, John Murray, London 1859.

2. Ch. Darwin, The Descent of Man and Selection in Relation to Sex, John Murray, London 1871.

3. G.H. Kieffer. *Bioethics* Addison-Wesley, Reading, 1979.

4. D.J Morris. *The Naked Ape.* Cape, London 1967.

5. F.J. Ayala. *Origen y evolución del hombre.* Alianza Editorial, Madrid 1980.

6. Congregation for the Doctrine of the Faith. *Dignitas Personae.* 8th Dec 2008. P. 5.

7. C.Rubbia. En Edgarda Ferri, *La tentazione di credere. Inchiesta sulla fede.* Rizzoli, Milán 1987.

8. J. Ayala. *Origen y evolución del hombre.* Alianza Editorial, Madrid 1980.

9. The International Human Genome Mapping Consortium "A physical map of the human genome", in *Nature* 409 (2001), pp. 934–941.

10. The Celera Genomics Sequencing Team. "The sequence of the human genome", in *Science* (2001), pp. 1304–1351.

11. F. Collins, E. Green, A. Guttmacher, M. Guyer, "A Vision for the Future of Genomics Research. A blueprint for the genomic era", in *Nature* 422 (2003), pp. 835–847.

12. F. Collins, M. Morgan, A. Patrinos, "The Human Genome Project: Lessons from Large-Scale Biology", in *Science* 300 (2003), pp. 286–290.

13. D. Reich, N. Patterson, M. Kircher, F. Delfin, MR. Nandineni, I. Pugach, AM. Ko, YC. Ko, TA. Jinam, ME. Phipps, N. Saitou, A. Wollstein, M. Kayser, S. Pääbo, M. Stoneking. "Denisova admixture and the first modern human dispersals into Southeast Asia and Oceania". *Am J Hum Genet.* (2011) Oct 7;89(4):516–28.

14. L. Abi-Rached P. Parham *et al.* "The Shaping of Modern Human Immune Systems by Multiregional Admixture with Archaic Human". *Science* (2011) 334 (6052).

15. The Chimpanzee Sequencing and Analysis Consortium. "Initial sequence of the chimpanzee genome and comparison with the human genome" *Nature* 437, 69–87 (1 September 2005).

16. D.P. Lodke *et al.* "Comparative and demographic analysis of orangutan genomes". *Nature* (2011) Jan 27;469(7331):529–33.

17. A. Scally *et al.* "Insights into hominid evolution from the gorilla genome sequence" *Nature* 483 (2012): 169–175.
18. F.C. Chen, W.H. Li. "Genomic divergences between humans and other hominoids and the effective population size of the common ancestor of humans and chimpanzees". *Am J Hum Genet* 68 (2001). (2): 444–456. doi:10.1086/318206.
19. T. Giger, P. Khaitovich, M. Somel, A. Lorenc, E. Lizano, L. Harris, M. Ryan, M. Lan, M. Wayland, S. Bahn, S. Pääbo, S. "Evolution of neuronal and endothelial transcriptomes in primates". *Genome Biology and Evolution* 2 (2010): 284–292.
20. K.S. Pollard, S.R. Salama, B. King, A.D. Kern, T. Dreszer, S. Katzman, A. Siepel, J.S. Pedersen, G. Bejerano, R. Baertsch, K.R. Rosenbloom, J. Kent, D. Haussler. "Forces shaping the fastest evolving regions in the human genome". *PLoS Genet.* 2 (2006) (10): e168.
21. S. Ptak, W. Enard, V. Wiebe, I. Hellmann, J. Krause, M. Lachmann, S. Pääbo, "Linkage disequilibrium extends across putative selected sites in FOXP2". *Mol. Biol. Evol.* (2009) doi:10.1093/molbev/msp143.

Francis Sellers Collins (born April 14, 1950)

12. On the Evolution Controversy[1]

Thomas B. Fowler

Darwin published his great work, *Origin of Species*, in 1859. There he propounded his notion of organic evolution, according to which all life forms arose by variations from a primitive ancestor. Darwin's key contribution was not the idea of organic evolution, but rather the naturalistic mechanisms that could make it work — natural selection together with genetic variation.[2] In Darwin's formula, the mechanisms give rise to *common descent with modifications*: a primordial living organism arose and replicated. Its descendants with slightly better (or much better) characteristics were able to leave more descendants than others (survival of the fittest). In this way there was a gradual improvement in the population, and eventually the improvements led to new species and higher taxa. The result was a gradual increase of complexity in flora and fauna, entirely mediated by natural processes, leading to the generation of all life forms. Though problems with the theory were quickly uncovered, it gradually won over the scientific community. However, by the early 20th century, many difficulties had accumulated, and new scientific results in the area of genetics were growing. An overhaul of the theory was therefore required. This was done in the first half of the 20th century, and work has continued to the present day. This new theory, or *New Synthesis* — more commonly known as *Neo-Darwinism* — is the school that now dominates the scientific landscape. Despite a number of serious theoretical and empirical shortcomings, this theory has the allegiance of the majority of scientists, educators, and educated people. Though much of the evidence for Darwinian evolution is perforce historical or circumstantial, and relies on inference rather than direct demonstration, its proponents believe that all of its problems will sooner or later be resolved, and that the evidence for the theory is so overwhelming that no serious questioning of it is possible.

This view, however, is not shared by everyone who is knowledgeable about the subject, and over the last forty years, the Neo-Darwinian school has increasingly found itself under attack from both scientists and non-scientists alike, especially over the idea that its proposed engine can account for all life.[3] Nonetheless the Neo-Darwinian school has maintained its hegemony in the academic and intellectual realms, and has found the courts sympathetic to its message that dissenting opinions should be suppressed, at least in the classroom. Today we have three other major theories of the history and development of life on earth, and at least seven important philosophical/religious interpretations of it. These are not independent, and in fact metaphysical assumptions are key drivers of both the scientific theories and the interpretations.

The other three major schools of thought about evolution are (1) *Creationism*, which accepts only a literal reading of the Bible, specifically Genesis 1, and consequently a young age for the earth; (2) *Meta-Darwinism*, which accepts naturalism but believes that the mechanisms proposed by the Neo-Darwinian school are inadequate to explain observed facts; and (3) the *Intelligent Design* school, which also rejects the adequacy of the Neo-Darwinian mechanisms, and in addition questions whether naturalistic mechanisms could ever account for observed complexity in life forms.

Surprisingly, many key terms related to evolution and the evolution controversy are poorly defined or have multiple meanings. Often the meaning of a term is changed in mid-argument. "Evolution" itself has many definitions: the simplest is just change over time, accepted by all schools. The evolution controversy turns on other meanings, which involve some form of *explanation* of change over time. The situation can be understood in terms of evolution levels, of which there are three:

1. Historical evolution: sequence of life forms over earth's history [change over time]. Basically the geological time scale, and the sequence of flora and fauna seen in the fossil record.
2. Common descent evolution: hypothesis that later life forms related to and arose from earlier ones, ultimately going back to single progenitor, in order to explain similarities observed in (1).
3. Strong Darwinian evolution: hypothesis that common descent and all observed changes in life forms can be explained by naturalistic processes of random mutation coupled with natural selection.

These are not synonymous! Historical evolution is a necessary but not sufficient condition for common descent evolution, and it in turn is a necessary but not sufficient condition for strong Darwinian evolution. Note that the evidence required to establish each of these levels is different, but the reader will rarely encounter a book or lecture where this fact is emphasized. Rather, evidence for the first is commonly used as evidence for the second and third, even though this is logically incorrect.

The main observable facts (i.e., things observable by anyone today) to be explained include:

- Traits of organisms and adaptations to their environment
- Fossils and the similarity of certain fossils to earlier and later forms
- Variation and diversity in populations
- Physiological and developmental similarities

- Genetic make-up, and genetic similarities
- The nature of living species
- The geologic column and its associated fossils

All science of the history of life must start from these facts and ultimately return to explain them.

The key issue in the evolution controversy is really easy to understand. Contrary to popular belief, it is not the existence of natural selection (survival of the fittest); that is accepted by all schools and never has been in dispute. Nor is it the fact of random mutations; everyone agrees that such mutations occur. The schools differ on question of the *extent to which these two processes can work together to create useful new genetic information and disseminate it in a population.* This new information, if it indeed is created, *must cause formation of new or better structures and systems that lead to viable organisms with improved functionality.* As it is often expressed, all schools agree that the Darwinian mechanisms can account for *microevolution* (small adjustments in characteristics of a population); they disagree about the degree to which they can account for *macroevolution* (creation of new, more advanced types of organisms), and ultimately new, higher taxa.

Neo-Darwinian school

The Neo-Darwinian theory is supported by all major scientific faculty and organizations. It is ubiquitous in textbooks at all levels, and regarded as established fact by the majority educated people, whether religious or non-religious (though as noted above, most erroneously assume that historical evolution is equivalent to Strong Darwinian evolution with respect to truth value and evidence). There are two essential hypotheses in the theory: (1) Common descent

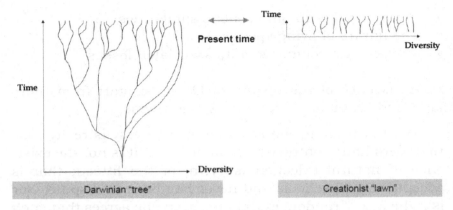

Figure 1. Comparison of Neo-Darwinian and creationist views of species formation.

— all organisms arose from a single ancestor at some remote point in past (see Figure 1); (2) Gradual improvement through incorporation of beneficial mutations. Specifically, gradual improvement in *phenotype* occurs through occasional small beneficial changes to *genetic* material that arise from random mutations. Resulting organisms are better suited to environment, and thus have greater chance of survival and propagation. The changes are incorporated into population's genetic reservoir by natural selection. Over many generations gradual improvements lead to emergence of novel traits and structures, and eventually results in new species and higher taxa (macroevolution). This is termed "phyletic gradualism". The process is shown as a feedback loop in Figure 2.

Obviously this theory makes many assumptions. The principal assumptions are (1) organisms can be continually improved by a series of small steps; (2) more favored varieties are more likely to survive long enough to breed; (3) there will be more of favored variety in next generation (natural selection); and (4) species will tend to evolve (improve) over

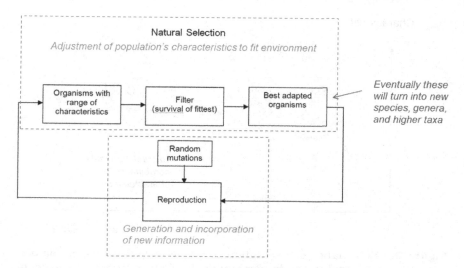

Figure 2. How the Neo-Darwinian pieces fit together.

time. But in fact additional assumptions are needed to make the scheme work: (5) there is a viable path of small variations connecting any two species (or other taxa), i.e., you can get new *architectures* by a series of *small changes* (no jumps!), as shown in Figure 3; (6) naturally occurring random processes can supply any new information needed for variations to traverse these paths; (7) it is possible to traverse the connecting path in finite time. The Neo-Darwinian school does *not* maintain that arbitrary transformations are possible, e.g., turning a dog into a cat. But it does maintain that any structure or system can, in a finite number of steps of finite probability,[4] be reached from another, *or from some predecessor common to them both.* In information theory language, the theory claims that random mutation and natural selection can create useful new biological information and incorporate it so as to produce viable new and improved organisms. *The Neo-Darwinian school has to deliver on this issue!*

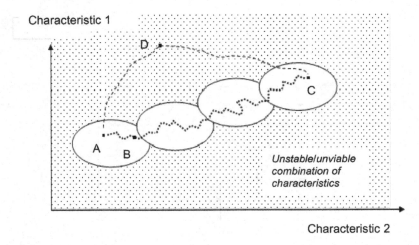

Figure 3. Schematic representation of stable or viable regions for biological entities. Stable/viable combinations of characteristics shown in white; other combinations are unstable. Neo-Darwinian theory assumes that no gaps exist between A and C, as shown; but the path through D would not be viable.

There is much evidence in favor of the Neo-Darwinian paradigm, mostly circumstantial but still powerful:

- Commonality of genetic code
- Similarity of biological and biochemical systems, structures reflected in classification systems
- Similarity of physiological characteristics
- Key genetic characteristics such as synteny blocks
- Development of antibiotic resistance
- Existence of possible precursors for some structural proteins

But there are a few problems. (1) The fossil record is really not in accord with the theory, showing jumps not gradual transitions. (2) Most evidence cited applies to common descent or action of natural selection, not strong Darwinian evolution, and thus is not germane to the heart of the dispute. (3) The Cambrian Explosion[5] indicates that higher taxa, including phyla, originated early in the history of life, rather than later, suggesting a top-down rather than bottom-up model as required by Darwinian theory. (4) There is no solid proof that random mutations can generate new information and new, higher forms of life. (5) Pleiotropy (multiple traits controlled by single gene) suggests that most areas where change is needed for evolution cannot change because of destabilizing effects. (6) Detailed mathematical calculations reveal problems; among them, it is unclear that incremental search (i.e., generation of random mutations) can solve the required non-convex high-dimensional optimization problem. (7) The necessary simultaneous changes in different systems required to create viable improved organisms have exceedingly low probability.

Summarizing the pros and cons for the Neo-Darwinian school, we have for the pros: (1) Fully naturalistic explanation to account for development and characteristics of life; (2) Based on simple, plausible mechanisms; (3) Supported by much empirical (albeit circumstantial) evidence. For the cons, we have (1) Does not account well for some evidence; (2) Glosses problem of need for multiple simultaneous changes; (3) Has not demonstrated that random changes can yield new information and improved organisms; and (4) Relies on long-range extrapolation.

Creationist school

By the early 1960s, the materialistic and frankly anti-theistic implications of Neo-Darwinism were setting off alarm bells in the evangelical community. Coupled with a commitment to some form of Biblical literalism, this led to a resurgence of *Young-Earth Creationism* (belief in recent creation of the earth, less than 10,000 years ago). Most early Creationist efforts tended to be rather amateurish, but over the years

quality has grown steadily. With increasing numbers of converts, especially among scientists, some modern Creationist
research and analysis is fairly sophisticated, if limited in
quantity. Of course, for the Neo-Darwinians and most other
scientists, the whole idea is laughable, and no amount of
sophistication can change that. Ridicule has never daunted
the Creationists, however, and they began preaching their
message in many venues, including schools and colleges,
where they challenged evolutionists to debate the issues.
They found sympathetic audiences in many places and by
the 1980s had become formidable advocates, winning over
many and often defeating opponents, who seemed confused
and bewildered by their arguments. This, understandably,
alarmed evolutionists, as did their efforts to win a place for
their views in public schools and textbooks. Those efforts
have triggered an on-going series of legal and legislative
battles. Creationist theories include the following points:

- Earth is fairly young (~10,000 years)
- There was a major flood which accounts for earth's geologic features
- Earth was repopulated after the flood
- Flora and fauna in fossil record existed contemporane-
  ously inhabiting different ecosystems, rather than different time periods
- Common descent occurs, but only from created kinds, so
  that most observed similarities are due to *common design
  plan* rather than common descent
- Evolution occurs, but is always degenerative

Several major problems confront Creationist theories.
(1) Distant starlight. Even a backyard telescope can see galaxies known to be millions of light years away, which means
that it took their light that long to reach us. (2) Geological
evidence of an old earth, e.g., eroded mountain ranges.
(3) Radiometric dating methods, which give ages of hundreds

of millions or billions of years for certain earth rocks. (4) The fossil record, though discontinuous, shows fossils embedded in rock millions of years old. (5) Uniformity of the genetic code and commonality of many genes, as well as physiological and biochemical similarities of flora and fauna, all of which point to common descent.

The make-or-break issue for the Creationists is the age of the earth. A short age for earth demolishes any evolution theory and establishes Creationism as only alternative; whereas a long age for earth demolishes Creationist science and theology. *The Creationists have to deliver on this issue!*

The Meta-Darwinian school

In 1966 a famous conference was held at the Wistar Institute in Philadelphia, *Mathematical Challenges to the Neo-Darwinian Theory of Evolution*. It was an outgrowth of an informal meeting at the house of physicist Victor Weiskopf and was chaired by biologist and Nobel laureate Sir Peter Medawar. The general thrust of the conference was that evolution is impossible if it must rely upon the mechanisms proposed by Neo-Darwinism, though by no means did all of the participants accept that conclusion.

There was also growing concern among biologists that explanations of matters such as discontinuities in the fossil record were not forthcoming from the Neo-Darwinian school. In response to this situation, which was not only embarrassing but also supplying Creationists with much ammunition, Steven J. Gould and Niles Eldridge formulated a new version of evolutionary theory, which they christened "Punctuated Equilibrium". This theory proposed that evolution happens in spurts, not continually and gradually, as demanded by the Neo-Darwinian model. Gould and Eldridge stirred up much opposition, and many felt that a summit conference was needed to deal with some crucial issues.

As a result, a conference was held at the Field Museum in Chicago in October of 1980, attended by the major figures in evolutionary biology. The central question addressed was the validity of extrapolation from micro- to macroevolution. There were considerable fireworks at the conference, and even *Science* magazine conceded that most felt the answer was "No". But by this time the influence of Creationism had become such a threat that, in an unprecedented move, no formal record was made of the conference proceedings, so as not to give Creationists aid and comfort.

So by the early 1980s, science had amassed a huge amount of circumstantial evidence pointing to evolution,[6] but some stubborn problems persisted. Hence scientists found themselves staring down a wrenching dilemma. On the one hand, many felt that the reigning Neo-Darwinian theory was seriously deficient, and needed work. However, they did not want to abandon the idea of naturalistic evolution. On the other hand, they felt (correctly) that any perceived crack in the scientific position, any suggestion that there was not monolithic agreement about evolution, would be ruthlessly exploited by Creationists. As most regarded the latter as the worst of the two alternatives, they tended to close ranks around the Neo-Darwinian theory, despite its admitted problems.

Some scientists, including Stuart Kauffman and Lynn Margulis, proposed new mechanisms for evolution. Kauffman emphasized the self-organizing qualities of matter as a way to deal with the long-standing problem of astronomically improbable events required by the orthodox theory. Margulis sought to defuse the problem by arguing that complex organisms can be built via symbiosis. Another dissenter, physicist and astronomer Fred Hoyle, even proposed that microscopic life originated elsewhere in the universe, where conditions might be more favorable. Their work, together with that of Gould and Eldridge and others constitute an eclectic cadre of theories forming the *Meta-Darwinian school,* the third major school in the evolution debate. This school accepts that life forms came about through naturalistic means, but believes that mechanisms beyond those of the Neo-Darwinian school are required to account for it. For individual biologists, the boundary between it and the Neo-Darwinian school is often not hard-and-fast.

The major issue for the Meta-Darwinian school is: Can other naturalistic mechanisms better account for history and development of life on earth than simple natural selection coupled with random mutation? *The Meta-Darwinian school has to deliver on this issue!*

Intelligent design school

By the late 80s an assorted group of thinkers had become quite dissatisfied with Neo-Darwinism, and in particular with the way it was being used to promote philosophical and anti-religious agendas. Though they drew on many of the same concerns and arguments as the Meta-Darwinian school, and to some extent the Creationists, they felt that a new approach was needed because of disagreements with these two schools. In particular, this group rejected the purely naturalistic outlook of the Meta-Darwinian school, but did not wish to embrace Creationist views on the young age of the earth, with its attendant implications for geology, physics, astronomy, and other sciences. Spearheading this group was Philip Johnson, a law professor who had studied evolution in depth. Johnson published several very influential books in the early 90s, beginning with *Darwin on Trial* (1991).[7] This book may be considered to mark the beginning of a new movement referred to by its members as the *Intelligent Design school,* though its roots can be traced back to work by Michael Denton,[8] Norman Macbeth,[9] the Wistar Conference,[10] Harold Yockey,[11] and perhaps even to the work of longtime critic of Neo-Darwinism Fred Hoyle.[12] Contrary to popular belief, Intelligent Design does not start with the idea of a "Designer"; the "Designer" is an extra-scientific inference from a particular scientific conclusion about the impossibility of certain physical transformations. In the same way, atheism is an extra-scientific inference from Neo-Darwinian theory drawn by many of its advocates, including Dawkins, Dennett, and Provine.

The Intelligent Design School *seeks to put the question of the ability of unguided natural processes to create new information on a rigorous footing.* It examines the creation of complex structures and processes, utilizing new concepts such as "irreducible complexity" and the "design filter". The School is really posing questions that should have been asked all along, though they are not questions that Neo-Darwinian school wants people to be asking. The basic argument of the ID school is that the complexity of biological structures and systems is such that there is essentially zero probability that they could have evolved even over billions of years with the random mechanisms at the heart of Neo-Darwinian theory.

Michael Behe's notion of irreducible complexity is based on a key fact of many complex systems, namely that they have at their core a set of interacting elements *all of which must be present and functional for the system to work at all.* Thus in the case of a watch, there must be a mainspring, appropriate gears to drive the hands, an escapement mechanism, and numerous other components. If any of these is missing or non-functional, the watch will not keep time. It is important to note that in such case, the watch does not retain *some* of its ability to keep time; it loses *all* of its ability to do so. This broken configuration serves no useful function and must either be repaired or discarded.[13] Behe himself uses the example of a mousetrap.[14] This situation, in which no component can be removed without destroying the functionality of the overall system or device is *irreducible complexity.*

The implication of irreducible complexity for biological systems is straightforward: such systems could not have originated through any incremental, step-by-step process, because *all* of the components must be present for the system to work; there is no partial functionality. Such an incremental change pattern is at the heart of Neo-Darwinian theory, since that theory has no alternative to create structures and systems except by small, incremental changes. Hence the irreducibly complex structures which Behe claims are common in biological systems cannot have originated as Neo-Darwinism claims, and therefore the theory must be false.[15] This is a strictly scientific argument, albeit one which rests on a crucial notion, that of irreducible complexity. Clearly the argument is sound only if Behe can demonstrate that some biological systems are, in fact, irreducibly complex. William Dembski has elaborated Behe's argument with a new concept, the *Design Filter*, which gives a more empirically-based method for determining if something can be attributed to random processes. If Intelligent Design is correct, it should become possible to map these gaps, and to understand them from a theoretical standpoint, that of physical chemistry. Such an investigation will likely involve a great deal of higher mathematics; we would have to examine biological structures in some very high dimensional spaces and investigate the effect of physical laws on their possible transformation in those spaces. The quantitative methodology to analyze complex structures and discern those which cannot have arisen by random processes is only the first step in building the new theory of possible biological transformations. In effect, Intelligent Design would reveal a deeper level of structural characteristics of biological systems and entities than is now recognized. If brought to fruition, this would have an impact not only on the study of evolution,

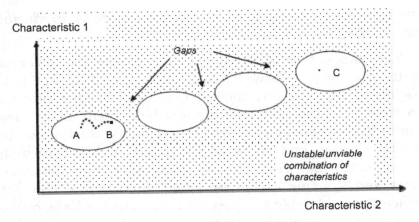

Figure 4. The ID school believes that the complexity of organisms and structures is so great that no viable path exists between some combinations of characteristics.

but on much modern-day research, because it would show that certain types of change cannot occur, while other types may be quite likely. The impact in areas such as development of antibiotics and antiviral drugs would be significant.

The key issue for the Intelligent Design School is this: Can restrictions on generation of new information (and thus on system, process and structure complexity) by random biological mechanisms be demonstrated scientifically? i.e., can ID show that the above-mentioned gaps exist? *The Intelligent Design School has to deliver on this issue!*

The pros for the ID school are (1) Have posed a critical question for evolutionary thought in a scientific manner; (2) Devising rigorous tests to determine if type of transitions required for naturalistic evolution can occur. The cons include (1) Have not definitively shown that transitions cannot occur; (2) Have not delineated types of permitted and forbidden transitions; and (3) Excessive reliance on probability calculations with questionable basis.

So the current situation with respect to science may be summarized as follows: there are four competing schools of thought about evolution. The principal divide is between those that believe natural mechanisms alone are sufficient to account for all life on earth (the dominant Neo-Darwinian school and the much smaller Meta-Darwinian school), and those that reject this belief (the Creationist school and the Intelligent Design school). The Neo-Darwinian school argues the sufficiency of random mutation and natural selection acting together to create the "common descent with modifications" paradigm. The Meta-Darwinians agree that natural mechanisms alone are sufficient to account for all life, but rejects the "one size fits all" approach of the Neo-Darwinian school, arguing that the mechanisms proposed by the Neo-Darwinian school are inadequate for the job, and are capable of explaining only a fraction of life's history. Another great divide is the age of the earth. Creationists dispute the old age claimed for the earth, and thus most of the conclusions of the Neo-Darwinian school about how life changed, including common descent from a single ancestor.[16] The Intelligent Design school accepts an old age for the earth, and all of the science behind it, and accepts much of common descent; but it rejects the sufficiency of natural mechanisms to account for all of the changes required for the history and development of life on earth. In terms of numbers and adherents, the Neo-Darwinian school has the allegiance of most scientists, educators, and college-trained professionals; the Meta-Darwinian school is very small but growing among those same groups; Creationism has made great inroads in the general public, where it far outweighs the other schools. Intelligent Design is still rather small in numbers but is gaining support among all groups.

There is no "Intelligent Design Science," or "Creation Science," or for that matter, "Neo-Darwinian Science." There is only science — good science and bad science, perhaps, but only science. Science is the pursuit of truth about nature. There cannot be two truths or Averroistic dual truth — there can only be two competing theories. One will eventually win out, or both will be overthrown. The competing theories can advocate different research programs, and be looking for different phenomena. Sooner or later the predictions of one will be vindicated, and those of the other will not. Intelligent Design proponents are right to seek for places where their theory can be tested and verified, just as Creationists are right to seek for those places where their theory of rapid, high-energy processes can be verified.[17] No one should be afraid of this competition, nor try to block it. Indeed, such challenges should be welcomed by any reigning theory, as they give the theory a wonderful opportunity to prove its superiority — if indeed it is the best theory. Nor is science a popularity contest; it is based on evidence and explanatory power. Just because there are competing theories doesn't mean that one can chose whatever theory one likes, fits into one's preconceived views, or is most popular.

Philosophical/religious interpretations of evolution

Many philosophical and religious interpretations of evolution have arisen because of its implications for man and his position in the universe. It quickly becomes clear to any objective student of the evolution controversy that one's position on evolution comprises two elements: a scientific judgment, and a philosophical position or worldview. These two are usually mixed together rather indiscriminately, and the resulting mix is often held passionately and defended tenaciously. This is conducive neither to good science nor to clear philosophical and theological understanding. Unfortunately few are sufficiently interested in the truth to consider their own views or those of their opponents in an objective manner. Pronouncements such as the following show just how easily it is to inflame passions:

> Evolution is promoted by its practitioners as more than mere science. Evolution is promulgated as an ideology, a secular religion — a full-fledged alternative to Christianity, with meaning and morality. I am an ardent evolutionist and an ex-Christian, but I must admit that in this one complaint — and Mr [Duane] Gish is but one of many to make it — the literalists [Creationists] are absolutely right. *Evolution is a religion. This was true of evolution in the beginning, and it is true of evolution still today... Evolution therefore came into being as a kind of secular ideology, an explicit substitute for Christianity"*[18] [italics added].

The seven major interpretive positions on evolution can be visualized as a spectrum, with atheism on one end and Creationism on the other, and the varieties of theistic evolution in the middle, along with other views tending to one side or the other, as shown in Figure 5. Curiously, the

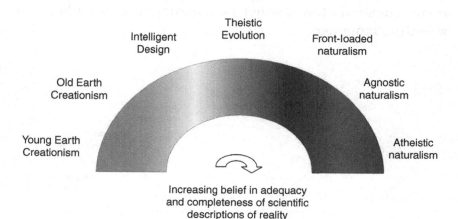

Increasing belief in adequacy
and completeness of scientific
descriptions of reality

Figure 5.

extreme positions, while diametrically opposite in some ways, share some fundamental deep-seated (and widespread) assumptions about the nature of knowledge. They assume that knowledge is monolithic in the sense that all "facts" are on the same level, specifically, the direct observational level. Thus contradiction between religion and science can — and does — occur at many points, at least in their view. Theistic evolution eschews such confrontation, and in its various forms always starts from a different assumption, namely that knowledge is multi-tiered, and "facts" cannot all be assumed to be on the same plane, much like the movement of airplanes in the sky: two airplanes may seem about to collide, when viewed by an observer looking straight up, though in fact they are flying at different altitudes. So basically accepting empirical science at the direct observational level, theistic evolution argues that (1) matter was created with the power to engender life and (2) we must step back and recognize the essentially hierarchical nature of human knowledge, with the direct observational level forming only part of it; philosophical and religious knowledge lies deeper, manifested by the many questions that cannot be meaningfully formulated in scientific language.

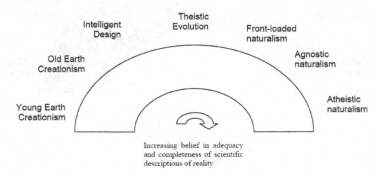

Figure 6. Spectrum of interpretations of evolution

Within most of these designations there is a range of interpretation. The positions can be summarized briefly as follows:

Young earth creationism

Events in Genesis are literally true, so that creation of the earth and the universe occurred in six 24-hour earth days. Any science that deviates from this account is either completely wrong, or represents a false interpretation of the empirical data. Since both science and Genesis are considered to be literally true, direct contradiction is possible on virtually all matters. The Creationists feel that generally accepted science (geology, astronomy, evolutionary biology) contradicts Scripture in three important areas: (1) the time frame of creation; (2) the order of creation; and (3) the manner of creation. They believe that their own version of science better accords with the observable facts.[19] While Creationists agree that species can change over time, and that new species can arise, they believe that any such change can only be neutral or degrade the genomic information of the species. Upward change is not possible. This reinforces the need to assume direct creative action by God to establish the original "kinds". However, Creationists do agree that natural selection can work on populations to optimize them for a particular environment, and interpret this as God endowing his creatures with the capabilities that they need to thrive. Despite some serious problems with sciences such as geology, physics, and astronomy, Young-Earth Creationism enjoys considerable public support in the United States and some other countries.

Old earth creationism

Earth is old, as astronomy, geology, and other sciences claim. Genesis 1 is not to be interpreted literally. There are two basic approaches to the theological issues, i.e., to reconciling the Bible with scientifically established time scales. First is the "progressive creationism" theory, according to which each "day" of creation can correspond to a very long period, and the days may even overlap. Each day corresponds to a specific aspect of God's creative work. The key theological message is that Divine action was required to cause the events narrated; random processes are not adequate. Time scales are unimportant for this message. But because Divine creative action is needed, Darwinian evolution cannot be correct as an explanation for the history of life, or at least for most of it. The second old earth creationist approach is the so-called "gap" theory, according to which there is a lengthy gap, perhaps billions of years, between Genesis 1:1 and Genesis 1:2. The gap theory argues that the earth became corrupted during that period, and had to be repaired by God as described by the six days of creation.[20]

Intelligent design

Virtually all of modern science is correct, especially physical sciences such as astronomy and geology. Common descent may also be correct, as well as historical evolution. The Intelligent Design school questions whether the type of transformation required by Darwinian evolution can occur at all, or within the time spans allotted, and they have sought to give empirically testable methods to buttress their claims. This is, of course, argument on the direct observational level. Claims about the feasibility of the transformations required by the Darwinian and Neo-Darwinian theories are strictly scientific question, but they have clear theological implications. If the transformations cannot occur as required by the Darwinian and Neo-Darwinian theories, then some external source must be required to bring them about. It is this conclusion — that naturalism ultimately fails — which causes such consternation among proponents of Neo-Darwinism, because they realize all too well the extra-scientific implications. The ID school thus accepts that matter is endowed with great power, but not all the power that Darwinism needs for a completely naturalistic explanation of the history of life.

Theistic evolution

Theistic evolution accepts science at face value, and maintains that the scientist will never encounter any sort of "wall" blocking his progress, such as that claimed by the Intelligent Design school. However, it argues that theologically this is not important because the range of questions that science can meaningfully ask and try to answer is not coextensive with all of human knowledge, and in particular, does not even cover much of what we do in our daily life, especially with respect to intentions and morality. The fact that we even talk about science, knowledge, and truth indicates that there is philosophical knowledge that cannot fall within the scope of science. Therefore contradictions between science and religion, properly understood, are illusory.

Rather than maintaining such a direct challenge to empirical laws, theistic evolution argues that we must step back and look at human knowledge as a whole. That is, we must recognize its essentially hierarchical nature, with knowledge at the direct observational level forming only part of it. Thus what the scientist finds is not the whole of reality or the whole explanation of reality. In other words, there are aspects of reality that are not accessible to science, or even meaningfully describable in scientific terms. It is here that theology has its meaning (and philosophy as well). So both theology and philosophy operate on a deeper level than the level of phenomenal appearances, the presumed realm of empirical science. For example, consider a theological doctrine such as Divine creation of the universe. For the theistic evolution proponent, no theory about this contradicts science because science only investigates phenomena, and not things such as creation *ex nihilo*.

Front-loaded naturalism

This theory, hearkening back at least to St. Gregory of Nyssa (c. 335–394), asserts that while things happen according to what science says, and evolution unfolded in the naturalistic way envisioned by Darwinists and Neo-Darwinists, in fact it was all pre-ordained because of the laws governing matter. In effect, man was the inevitable product of the forces set in motion at the time of the Big Bang. This is in contradiction to a commonly held idea about the unfolding of life promoted by Gould, among others, according to which much of life's unfolding was accidental, and small shifts might have resulted in completely different life forms today. While the idea that God took a "hands off" approach may seem to hark back to Deism, in fact it has a much more theological flavor, because of the fact that the emergence of man was essentially programmed in. This approach permits one to view evolution at the direct observational level while at the same time recognizing other, theological dimensions of it.

Agnostic naturalism

Agnostic naturalism accepts that science is correct. In particular, it accepts that Darwinian and Neo-Darwinian explanations of the history of life are correct. It stops short of claiming that science exhausts human knowledge, but does not accept that there is any truth to theological knowledge or valid claims to it. This position tends to shade into Deism. Its proponents have an uneasy feeling that science might not be the whole story, either with respect to knowledge or in ethical matters, but believe that religion, especially organized religion, has too many problems to be credible in any area.

Atheistic naturalism

Atheistic naturalism assumes that all knowledge is scientific, or at least all knowledge that has any degree of certainty and therefore of credibility. Though it concentrates on the direct observable level, it is ultimately based on the "unholy trinity" of naturalism, nominalism, and reductionism, as will be discussed below. In this view, which has a positivistic epistemology, all knowledge is on the same level, so any purported theological "knowledge" must be capable of being judged by empirical science, since that is the ultimate source of knowledge about the world. With respect to man, since Darwinian evolution can account for the history of life, it has eliminated the need for any Divine creative action. Therefore man is not special in any way, and the stories in the Bible are either false or merely edifying tales. Indeed, the history of life could easily have been different, with entirely different creatures, had circumstances been different. In this view ethics or morality tends to assume some form of utilitarianism because it seems to accord with the code of "survival of the fittest" at the population level.

Conclusion

A broad spectrum of philosophical and theological interpretations of evolution exists, ranging from Creationism to atheistic naturalism. All of these interpretations are built upon a worldview and a particular scientific theory of evolution, and these in turn rest upon a set of philosophical assumptions concerning epistemology and metaphysics.

Thus all interpretations of evolution make assumptions about the levels at which human knowledge exists. There is no science without philosophical assumptions, especially about the canon of reality and the scientific method. Of particular interest is the fact that some interpretations of evolution ignore (or pretend to ignore) the philosophical, concentrating on its favored scientific theory and worldview instead, both of which operate primarily at the level of direct observation. If one views all knowledge as being at that level, then collisions between theology and science are inevitable, especially in light of the fact that the Bible and other religious texts are not and were never intended as science texts. Such conflicts or contradictions between theology and science have led Creationists to reject many scientific conclusions and indeed entire theories, and atheistic naturalists to reject the possibility of theological knowledge. A better understanding of evolution comes from understanding the assumptions made at the lowest level. Some may choose to moderate or change their position when they recognize the assumptions they are making.

References

1. Parts of this chapter are based on "The Scientific Status of Intelligent Design", by Thomas Fowler, *Faith and Reason*, Vol. 31, No. 4, 2006, pp. 503–538; *The Evolution Controversy*, by Thomas Fowler and Daniel Kuebler, Baker Academic, 2007; and "Overview of the Theological and Religious Interpretations of Evolution", *Modern Age*, Vol. 52, No. 4 (Fall, 2010), pp. 271–284.
2. This is modern terminology; Darwin did not know about genes but surmised that there had to be some type of hereditary material subject to random changes.
3. David Stove, *Darwinian Fairytales*, New York: Encounter Books, 1995.

4. By "finite probability" here, we mean that the transition has a reasonable chance of occurring in an appropriate time span. Given that the age of the universe is estimated at 10^{18} seconds, a probability on the order of 10^{-30}, probably 10^{-15}, is not reasonable.

5. The Cambrian Explosion was a rapid development of life forms, including all major phyla, about 530 million years ago.

6. Virtually all of it supported what we have termed historical evolution; a lesser but still significant amount supported common descent evolution. Little if any supported strong Darwinian evolution, and this was the crux of the problem.

7. Phillip Johnson, *Darwin on Trial*, 2nd edition, Downers Grove, IL: Intervarsity Press, 1993. First edition published in 1991 by Regnery.

8. Michael Denton, *Evolution: A Theory in Crisis*, London: Burnett Books, 1985.

9. Norman Macbeth, *Darwin Retried*, Boston: Harvard Common Press, 1971.

10. Moorehead, P.S., and Kaplan, M.M., editors, *The Mathematical Challenges to the Neo-Darwinian Interpretation of Evolution*, Philadelphia: Wistar Institute of Anatomy and Biology, 1967.

11. Yockey, H.P., *Information Theory and Molecular Biology*, Cambridge University Press, 1992.

12. Hoyle, Fred, and Wickramasinghe, N.C., *Why Neo-Darwinism Does Not Work*, Cardiff: University College Cardiff Press, 1982.

13. An intelligent agent might be able to reuse some pieces from the broken watch to create another device; but that would be *intelligent design*, of course.

14. Michael Behe, "Darwin's Breakdown," in *Signs of Intelligence*, ed. By William Dembski and James Kushiner, Grand Rapids: Brazos Press, 2001, pp. 93ff.

15. They could, in theory, arise by "jumps" or "saltations", as they are called. But this idea, championed most recently by Richard Goldschmidt, has been resoundingly rejected by the Neo-Darwinian school in favor of small, incremental changes.

16. Contrary to popular belief, Creationists do not reject the notion of common descent, as they need it to account for repopulation of the earth after Noah's flood. They do reject the idea that all of life could have come from a single ancestor, since they believe that only degenerative change can occur naturally.

17. Creationists theorize that high energy processes acting over short periods (instead of the usual low energy processes acting over long periods) formed the principal geological features of the earth. For further details on their theory, see chapter 6 of *The Evolution Controversy*, by Fowler and Kuebler, Grand Rapids: Baker Academic, 2007.

18. Michael Ruse, "How Evolution Became A Religion: Creationists Correct?" National Post, p. B1, B3, B7 May 13, 2000.

19. There are different branches of Creationism, with differing theories.

20. *http://www.answersincreation.org/old.htm* (accessed 21 April 2009).

The Biochemical Challenge to Evolution

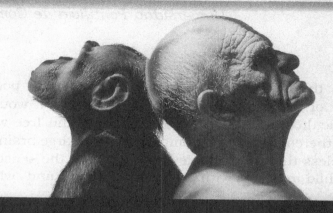

DARWIN'S
BLACK BOX

"No one can propose to defend Darwin without meeting the challenges set

out in this superbly written and compelling book."

—David Berlinski, author of A TOUR OF THE CALCULUS

MICHAEL J. BEHE

13. On the Riddle of Man's Origin

Manuel M. Carreira, SJ
Universidad Pontificia de Comillas

From the study of fossil bones, isolated from possible activities of the primates represented by them, it would be very difficult to scientifically infer intellect and free will or the lack thereof: there are animals with very large brains but nevertheless they show no desire to know in the sense that even a child indicates with a constant "what?" and "why?"

Instinctive behavior, genetically programmed, can include a process of "learning" by imitation as we find in the way birds "teach" their young to fly. Even a simple cleaning process (washing food to free it from sand) can be imitated and used by later generations without a conceptual transmission that would be true culture. This kind of learning is clearly the first step towards knowing what to do in most of our daily activities.

Intelligence shows itself mostly in the desire to know "useless things": why the Sun appears and disappears daily, what are the stars, why the Moon changes shape each month. In fact, Astronomy was the origin of modern science.

If a possible benefit — better chances of survival — were the reason why Man develops as a rational animal, we would expect evolution to have worked from Man to monkey. Biologically, a newborn chimpanzee is much better prepared to survive than a human baby, the most defenseless living thing for the longest period of time. Nature should have developed and favored the survival of the fittest by self-sufficiency in the shortest time possible.

An animal learns few things, but very quickly, and by instinct or imitation very soon is able to grow and function independently of the parents. Human development! needs a cultural stimulus even for the ability to speak and to think: a baby who lacked the attention of parents never develops adequately the brain to permit truly rational and intelligent behavior, the traits of a person who seeks Truth, Beauty and Good.

There is no way to explain rationality in terms of survival of the fittest, the reason given by merely evolutionary theories. Evolution occurs in biological-material structures: its effects can only be due to the properties of matter. But matter cannot be the reason for thought and free acts. The triple tendency previously mentioned, towards Truth, Beauty and Good, cannot be the result of electrical currents in the brain. Those are very similar in brains of animals totally devoid of intelligence and the freedom that constitutes Man as a subject of rights and duties. Nobody wants to be considered a robot ruled by physical necessity, and everybody expects recognition for scientific insights or heroic activities. If those who hold a purely materialistic view of Man fail to live according to their abstract tenets, they deserve to be ignored.

We find in Man an undeniable double level of functions that require a double source. We are part of the biological world, sharing components and processes with all other life forms on Earth, but it is still more obvious to each one of us that our thinking and free will set us apart from all other living things. Most important historical developments can be traced to "an idea whose time had come" and that changes the way society and personal relationships are structured. Nothing like that can be found in purely animal life.

Several cultures — in different places and at different times — have led to a variety of social structures, as clearly found in ancient history and also in our own time. From the prevalence of slavery, tribal rivalries and theocratic kingdoms, we have seen tyrannical materialism develop in communist countries, in a sharp modern contrast with democratic ideals. Mutual interdependence is the clear new message that prevails in international discussions of how to make the Earth livable for future generations, and how to provide goods and cultural opportunities for the entire human population.

Science gives us ever greater control upon natural resources, while — at the same time — increasing our ability to develop knowledge and awareness of far-reaching implications of that knowledge and its practical uses. On the other hand, even fast and universal sharing of knowledge can develop into a lack of privacy and, to some extent, a negative atmosphere in which news seem to stress problems, natural or man-made calamities, with an implicit rule that "the only news is bad news". A hidden materialism is behind an almost obsessive insistence upon Man's lack of special dignity within the natural environment, that leads to speak of "rights" of animals an even plants and mountains. It has reached the point of considering our presence on the planet as an infection that would be better suppressed! Hence plans to reduce birth rates, promoting as a service to the world campaigns of abortion, and finally euthanasia. We really have the absurd of presenting our evolution as justified to suppress mankind.

Earth is the common HOME, and we should appreciate it and care for it as the common good requires, including -of course- the good of coming generations. But creation would be pointless without leading to persons capable of a responsible response to the Creator, before whom every human being appears as a living image with a unique dignity that points to an eternal sharing of God's happiness. Without this transcendental horizon, the Universe would be absurd, as a series of material developments leading from initial chaos to final darkness and empty space. This is so stated by Nobel winner Steven Weinberg: "The more we know the Universe, the more absurd it seems".

A more logical way of thinking takes into account the "Anthropic Principle" proposed by scientists, not theologians or philosophers. It points to the very precise values of material properties, without which the existence of intelligent life would be impossible. A basic logic requires that those concrete values -not imposed by the very concept of matter- be chosen for a reason, when the Universe comes into existence without a previous state from which its parameters could be derived. The reason that most precisely appears as requiring an exact set of properties is the eventual appearance of intelligent life.

From the Big Bang almost 14 thousand million years ago, the evolution that gave rise to stars and galaxies led to the formation of a variety of planets around stars of different luminosity and evolutionary development. A very special star — the Sun — more massive than 92% of all stars, was born in a region of the Milky Way profiting from heavy element synthesis in previous generations, and safely far from deadly radiation from a central black hole. At the correct distance and with adequate mass, a massive Moon, and a dynamic structure leading to plate tectonics and a magnetic field, the Earth is the jewel of the solar system, the only place known where life could be sustained for billions of years. This is our home.

We have no scientific explanation for the first mystery, the step from inanimate matter to the first living cell: the information encoded in a single DNA molecule cannot be due to the chance arrangement of chemical bonds in the entire history of the Universe. Once life appears, we face its increasing complexity, the development of plants, the multitude of genetic changes *in many individuals simultaneously* that can give rise to different species. We should accept that evolution has taken place, but its process is far from clear when we ask how it actually happens. And any significant change in that evolutionary chain would have rendered impossible our existence.

Pre-human primates can be seen as preparing biological structures for the final step, that is the appearance of intelligence in Man. We thus reach now the second riddle for evolution, the most difficult one to answer in terms of biology, because it is no longer a question of development in degree, but of something totally new: the human spirit, intelligent and free.

Apollo 11 was the spaceflight which landed the first humans on the Moon on July 20, 1969.

14. On Science, History and Free Will

Lucía Guerra-Menéndez

Four luminaries of the 20th century who from different perspectives made important contributions to the philosophy and history of science were Pierre Duhem (1861–1916), Gilbert K. Chesterton (1874–1936), Étienne Gilson (1884–1978) and Stanley L. Jaki (1924–2009).

Duhem was a great French theoretical physicist, mathematician and historian of science who discovered the medieval origin of the concept of inertial motion in the works of Jean Buridan, (a teacher at the University of Paris in the 14th century), and wrote a fully documented account on the historical development of cosmological ideas: "*Le Système du Monde: Histoire des doctrines cosmologiques de Platonà Copernic*" (Paris, 1913–1959).

Étienne Gilson was a French historian of philosophy, specialized in mediaeval philosophy and in Thomas Aquinas. One of his greatest insights was the awareness that the philosophy of his generation was being absorbed by science.

Gilbert Keith Chesterton is one of the 20th century's greatest British authors and one who cultivated all literary genres: essays, narratives, biographies, poetry, journalism and travel books. He was also a great literary critic and, surprisingly, because he was a very humble man, with no pretensions of academic erudition, he was a true "seer of science". In 1922 he converted to Catholicism and his conversion was a shock for many of his friends and also of his critics. His cultural influence, very great when he was alive, is still felt today worldwide, especially in the English speaking countries, UK, US, Canada, Australia, etc. Among his most celebrated essays are: "Orthodoxy", "The Everlasting Man", "Saint Francis of Assisi" and "Saint Thomas Aquinas".

Stanley L. Jaki, Hungarian born Benedictine priest from the great Abbey of Pannonhalma, with a doctorate in theology from the Pontifical Atheneum of St. Anselm in Rome, and a doctorate in physics from Fordham University, N.Y., where he studied under Victor Hess, 1957 Physics Nobel Prize winner, the discoverers of cosmic rays. Jaki was a Gifford Lecturer at the University of Edinburgh (in 1974–75 and 1975–76) and a Fremantle Lecturer at Balliol College, Oxford (1977), awarded with the Lecomte du Noüy Prize in 1970 and the Templeton Prize in 1987.

Jaki was certainly one of the most original and penetrating philosophers and historians of science of the second half of the 20th century. His lectures were published under the title *The Road of Science and the Ways to God* (Chicago: University of Chicago Press; Edinburgh: Scottish Academic Press, 1978). He wrote over fifty books and hundreds of essays on the history of science, the philosophy of science, and on the relation between science and the Christian Faith.

In this Chapter we summarize and comment on Prof. S.L. Jaki's essays on history, science and free will from his book *Means to Message: A Treatise on Truth* (William B. Eerdmans: Grand Rapids, MI, 1999).

Science

Man stands out conspicuously[1] among all the higher animals because of such spectacular inventions as the making of *fire*, the planting of *seeds* to get food, the *domestication* of animals for various purposes (including transportation, food, clothing, helpful and defensive company, etc.) *writing* and *printing*. All of these inventions are themselves rudimentary steps in *technology*. But in all of them there is some kind of rudimentary *science* also.

Science could be ultimately characterized as a set of quantitative correlations of quantitatively correlated data of observation. It presupposes certain philosophical founda-tions rather than *providing* these foundations. For example, if there is not objective reality independent of the observer, there is no science. If identical physical observation at different times and different places were to give different results, science, would be impossible. Science assumes a coherent world to investigate, and a consistent capability in its observer, to begin with.

Around the beginning of the 16th century, a scientific approach to nature which had been already under way for three centuries in medieval Christendom began to make pro-gress at a quick pace. It accelerated in the 19th century, and finally exploded in the last century, at such a tremendous rate, that the promise of increasing, uninterrupted progress pervaded public opinion, giving the impression that any other cultural message, including philosophy and religion would be soon drowned out in science. Social scientists have made a habit lately of making science the *only* source of information worth considering. And this is applied to all areas of human concern: politics, family life, culture, enter-tainment and philosophy.

As S.L. Jaki notes,[2] if science's message was already becoming overwhelmingly dominant in the 19th century, because of the development of classical physics (mechanics, optics, chemistry, medicine, thermodynamics and electro-magnetism), much more so was it in the 20th, because of the development of modern physics (atomic and molecular, solid state and condensed matter physics; quantum chemistry, molecular biology, genetics, etc.). It has become a fashion, at least among those who know at most a little of "soft" science, to claim that today's science has the answers even to the last questions man asks himself, displacing therefore philosophy and even religion from the task.

Early in the 20th century, a prominent German theologian and historian, Adolf Harnack singled out Albert Einstein and Max Planck as the two greatest philosophers of modern time. While they deserve of course to be recognized as the two greatest 20th century physicists, as discoverers, respectively, of quantum physics and the theory of relativity (special and general), this does not make them great philosophers. They achieved their respective great discoveries as physicists, it is true, from a realistic perspective, equidistant as much from shallow empiricism as from "a priori" idealism, but that is all.

Renown as a scientist has become in our times a guarantee of deep philosophical perspicacity when coupled with an attractive literary style. For instance, the runaway commercial success of Jacques Monod's *Chance and Necessity* (1970) and Stephen Hawking's *A Brief History of Time* (1988) has been taken as proof that these two notable scientists are also competent philosophers. In particular, as noted by S.L. Jaki,[33] Hawking's book has been taken by some as nothing less than a synthesis of the philosophies of Kant and Aristotle, something which has been tried many times by professional philosophers all through the 20th century, unsuccessfully, because Kant is the epitome of subjectivism-relativism while Aristotle is the model of classical realism.

According to G.K. Chesterton, the word "Science" is used nowadays as an authority on everything: a vague authority that is often invoked but never truly pinned down. Everybody now looks to "Science" to solve all our problems.[44]

"What we have suffered from in the modern world is not...physical knowledge itself, but simply a stupid mistake about what physical knowledge is and what it can do. It is quite as obvious that physical knowledge may make a man comfortable as it is that there are such things as drugs but there are no such things as love-potions.

Physical Science is a thing on the outskirts of human life; adventurous, exciting, and essentially fanciful. It has nothing to do with the center of human life at all. Telephones, flying-ships, radium, the North Pole are not in the ultimate sense good, but neither are they bad. Physical Science is always one of two things; it is either a tool or a toy. A toy is a thing of far greater philosophical grandeur than a tool; for the very simple reason that a toy is valued for itself and a tool only for something else..."

History

Man must control his own nature, if he is willing to follow his higher aspirations, struggling with his lower instincts. Modern criminology is indisputable proof of that conflict whose depths modern psychoanalysis has failed to solve in a satisfactory way.

History[5] (political, cultural, social) would remain a riddle if that inner conflict between the higher aspirations and the lower instincts in man is not recognized as all important. According to Jaki *salvation history* including miracles–large or small–faces honestly that grim struggle, both within man and within society, which only utopians or social engineers are happy to ignore.

As presented in biblical revelation, the idea of God's Kingdom, both in the Old and the New Testament, but most clearly in the latter, is characterized by a *realistic view* of human history: the wheat and the tares going side by side within it until the end of time. Truly, God created the world and he saw that it was good, but he created man free, and man can chose from the beginning between good and evil. No rosy expectations are justified in this perspective of the Kingdom of God.

It is instructive,[6] according to S.L. Jaki, to take up the historical question of the origin of science from the view point of the great non-Western powers: the Muslim World, India and China. Why is it that China, the land which saw the first use of magnets and gunpowder, did not see the rise of a Galileo or a Newton? And what about India, which was able to produce rust-free iron two thousand years before the west, and invented phonetic writing independently of Egypt? Why did the Muslim world, only half a century ago, have to borrow from the West the technological know-how to harvest its vast oil fields? We find a double answer: first, referring to the ancient cultures mentioned above, all of them "came to a stop after making a few steps in the direction of the three laws of motion, basis of exact science, because they viewed the world as an eternal treadmill, doomed to endless returns after every Great Year. For them the status quo was the most that could ever be achieved. Christians by contrast believe in a Creation out of nothing and a single one-directional movement in time". Second, referring specially to the Muslim world, the answer may have something to do with the philosophical outlook favourable to freedom which blossomed in Catholic Medieval Europe at the time in which the great gothic cathedrals, the first universities and the first representative democratic institutions appeared in Christendom.

Shu King-Shen, Chinese ambassador to Saint Petersbourg[7] at the end of the 19th century said to his disciple Lou Tseng-Tsiang, future Chinese Minister of Foreign Affairs: "The strength of Europe is not in her armaments or in her science; it is in her religion ... Look at the Christian religion...when you have taken up its heart and its strength, carry them and give them to China". The disciple converted years later to Catholicism, and he opened the way in 1917 to establish diplomatic relations between China and the Vatican. But the European powers objected. Dom Lou died, January 15th 1949, in Bruges, Belgium, as abbot of the Benedictine monastery of St. Peter the Great. His fervent desire of bringing together the West and the Chinese Far East, heir of the wisdom of Confucius, met with insurmountable difficulties. But the Christian West was subject then, and it is now more so, to a tremendous wave of radical de-Christianization.

G.K. Chesterton, a seer of science[8] and also a seer of history[9] said in 1933 (*G.K.'s Weekly*, Nov. 2): "A society is in decay when common sense becomes uncommon". As a 20th century *prophet* he was astonishingly accurate:

He died three years before the beginning of World War II, but he had warned about Hitler before anyone else.[10]

He warned also very early about an outbreak of violence against the Jews.[11]

He predicted that the war would begin on the Polish border.[12]

He saw that the airplane would totally change warfare, extending the horrors of war to innocent civilians.[13]

He recognized that the new technologies, based upon the new 20th century science, could be used to improve ways of killing people.[14]

In the emerging industrial society, he said, electricity, water power, oil, etc., could be used to reduce the work of individual workers to a minimum, but he predicted also that machines would become their masters.

Chesterton also warned that technology would create as many problems as it solved: "The same industrial civilization, which aims at rapidity, also produces congestion".[15]

He predicted both the rise[16] and the fall[17] of communism in Russia. And also that the next great heresy would be an attack on morality especially sexual morality.[18]

He expected "a fashionable fatalism founded on Freud",[19] and he said that it would "exalt lust and forbid fertility".[20]

He predicted very early that abortion would be considered a sign of "progress".[21]

Chesterton is surprisingly on target when he says: "The true religion of today does not concern itself with dogmas and doctrines. Indeed it concerns itself almost entirely with diet."[22] And he adds: "Modern materialism... is solemn about sports because it has no other rites to solemnise".[23]

F. Fukuyama in *The End of History* proclaims[24] that the goal of history is a comfortable living that a technology supported market economy can make globally available. But we are seeing already in this second decade of the 21st century that Fukuyama's picture does not fit with reality. Is there any sign that rich nations are seriously interested in building up the economies of poor nations?

Free will

Free will is in man a very obvious thing. As noted by S.L. Jaki,[25] the great American physicist Arthur Holly Compton (1892–1962) Physics Nobel laureate in 1927, illustrated very graphically the indisputable reality of *free will*. He held that the ability one has to move or not his finger is something known to us much more directly and with much greater certainty than the well-known and well tested laws of Newton. If those laws were to deny his ability to move his finger at will, Newton's laws, according to Compton, would need to be modified.

René Descartes (1596–1650), a very clever geometer and speculative philosopher, who nevertheless contributed much to set Western philosophy on a wrong path for centuries, said[26] that it is so evident that we have free will...that it can be taken as one of our most common notions. However, abstract notions, as S.L. Jaki points out, are not amenable to quantitative measurement, and they tend to escape our conceptual grasp whenever we aim at "clear and distinct" definitions. In his early years, Descartes declared "Free Will" one of the three greatest miracles, "Creation" and "Incarnation" being the other two.

Baruch Spinoza (1632–77), a most consistent Cartesian, on the other hand,[27] expelled "free will" from the topics worthy of philosophical consideration. In his *"Ethica more geometricodemostrata"* he declared that the will was no different from the mind, which he took for a machine. When he said that the mind "is determined to wish this or that by a cause, which has also been determined by another cause, and this last by another cause, and so on to infinity..." he was clearly choosing deterministic pantheism as his own philosophy. He overlooked the dubious logic of his first step as well as that of the inconsistency of regression to infinity.

Immanuel Kant (1724–1804), as S.L. Jaki notes, laid a mine field[28] against his own system of critical reason when he radically questioned free will in one of the four antinomies of his *Critique of Pure Reason*. When he tried to say that it was unreasonable to talk about the soul as that agent that alone can be free, he was implying then and there that anybody talking about free will could do it only uncritically and wilfully, including, *himself.* The other alternative would be, of course, to talk about free will deterministically.

Existentialism, which recognizes only momentary acts, does not explain how those acts can be free. And much less how those acts can result in a free and wilful determination to pursue a given path to a given end for many years.

Prof. S.L. Jaki notes in his *Means to Message*:[29]

"Truth is a relevance of notions to objects that transcends the moment, including the always momentary registering of their reality. Insofar as that registering is limited to its being subjectively experienced, it is void of communicability and therefore may be left as a variant of solipsism. But truth is far more than what empiricists and physicalists (who do not recognize spiritual realities) can dream of, although they stake their case with the reality of physical objects".

In other words, free will proves false *mere material existence*. And G.K. Chesterton puts it into perspective:[30]

"The Humanist says to the Humanitarian: "You are always telling me to forget divine things and think of human rights. And then you talk me eagerly and earnestly about the pathetic helplessness of human beings, their faulty environment, their fatal heredity, their obvious animal origins, their uncontrollable animal instinct, ending with the old fatalistic cry that we must forgive everything because there's nothing to forgive. But these things are not the *human things*. These are specially and specifically the subhuman things; the things we share with nature and the animals. The specially and outstandingly human things are exactly the things that you dismiss as merely divine things. The human things are free will and responsibility and authority and self-denial, because they exist only in humanity."

In S. L. Jaki's reflections about science, history and the free will is stressed the authenticity of metaphysical thought, its object, which is the ontological reality of things, and the methodology to study them in the light of their ultimate causes, in order to be able to reach the ultimate cause, that is God. Given the claims of pseudo-scientism to reduce reality to the merely observable and measurable, Jaki reveals a world that exceeds the merely material determinism through a free will, which, precisely because it is free, escapes the narrow confines of the scientific laws and reveals the world of the human soul.

Finally I would like to write a few lines recalling Fr. Stanley Jaki as a person. I am very indebted to him for sharing with me his illuminating erudition and his warm personality in an admirable strong character full of vitality. I am grateful also for his example: *Oraetlabora*. It has been a great privilege to be working with him for years. God bless him. I am grateful for the opportunity of making this modest contribution to *Intelligible Design* edited by Julio Gonzalo and Fr. Carreira, and to FSJB (Father Stanley Jaki Brigade) for their support and constructive criticism.

Reference

1. Stanley L. Jaki, *Means to Message* (William B. Eerdmans: Grand Rapids, MI, 1999) pp. 43–62.
2. *Ibid.*, p. 43.
3. *Ibid.*, p. 44.
4. G.K. Chesterton, *Illustrated London News*, Oct. 9, 1909.
5. Stanley L. Jaki, *ibid.* p. 197.
6. *Ibid.*, p. 204.
7. See f.i. Fr. F. Lelotte, S.J., *La Antorcha Encendida* (Studium: Madrid, 1966).
8. Stanley L. Jaki, *Chesterton, A Seer of Science* (University of Illinois Press: Urbana and Chicago, 1986).
9. Dale Ahlquist, *G.K. Chesterton. The Apostle of Common Sense* (Ignatius Press: San Francisco, 2003). Chapter 15. Below we give the compact references as registered by Dale Ahlquist in this book.
10. I.L.N., Dec. 19, 1931.
11. I.L.N., Feb. 28, 1914; Sep. 4, 1920.
12. I.L.N., Dec. 2, 1933.
13. *The Outline of Sanity*, in *Collected Works* 33: 510.
14. *Ibid.*, 158.
15. I.L.N., March 21, 1925.
16. Introduction to Gorky's *Creatures That Once Were Men* (1905).
17. I.L.N., July 12, 1919.
18. G.K.'s Weekly, June 19, 1926.
19. I.L.N., May 29, 1920.

20. *The Well and the shallows*, C.W. 3: 501–2.

21. *Ibid.*, C.W. 3: 530.

22. I.L.N., May 11, 1929.

23 I.L.N., Nov. 15, 1930.

24. See f.i. Stanley L. Jaki, *Means to Message*, p. 203.

25. See f.i. Julio A. Gonzalo, *Dios y los Científicos* (Ciencia y Cultura, Madrid, 2006), pp. 39–4?.

26. "Les Principes de la Philosophie", I. 39, in *Ouvres de Descartes*, ed. C. Adam and P. Tannery (Paris: J. Urin, 1964) Vol. 8, p. 41.

27. See *The Chief Works of Benedict de Spinoza*, tr. R.H.M. Ewes (1883), (Dover: New York, n.d. Vol. 1).

28. See f.i. Stanley L. Jaki, *Means to Message*, p. 69.

29. *Ibid.*, p. 64–65.

30. G.K. Chesterton, *All I Survey* (Methuen and Co: London, 1934) p. 151.

GOD AND EVOLUTION

EDITED BY JAY RICHARDS

Concluding Remarks

According to George F. Smoot (Physics Nobel Laureate, 2006):

"Until the late 1910's humans were as ignorant of cosmic origins as they had ever been. Those who didn't take Genesis literally had no reason to believe that there had been a beginning".

It would have been perhaps more appropriate to say "taking Genesis seriously" than "taking it literally". But the implication is clear. In the ancient pre-Christian cultures as well as today in some modern de-Christianized environments, the world is seen as eternal, with no beginning and no end. That is why the concept of a cosmic first moment (a "big-bang") encountered strong apposition at its inception and was later on reluctantly reinterpreted in a cyclic or inflationary perspective.

Modern science, with its origin in the Medieval Christian West, developed vigorously during the last three centuries.

In the first half of the 20th century, revolutionary steps in physics, chemistry and biology — including quantum physics, relativity, nuclear, atomic and molecular physics, genetics and astrophysics — did prepare the grounds upon which the spectacular *technological revolution* we see today underway (nuclear reactors, solid state electronics, computers, artificial satellites, lasers, nanotechnology) is developing at an accelerated pace.

Contemporary men seem to be taking full control of nature's secrets and nature's potentialities. To a very large extent howeber, men are also beginning to forget that they have not created those impressive natural forces whose secrets previous generations of scientists had learned to understand and to control. However precisely because the Christian medieval wisdom which inspired the birth of modern science has been largely forgotten, the present technological revolution is getting out of control.

There are a few *singular moments* in cosmic history which point out to an intelligent and all powerful Creator:

About 13.3 × 10⁹ years ago: The formation of the first proto stars within the first proto-galaxies, a few hundred millions of years after the "big bang" (Our Sun is a star of second or third generation).

About 4.6 × 10⁹ years ago: the formation of the Earth-Moon system, at a privileged location in our solar system, evolving so as to develop a biosphere capable of sustaining life.

About 3.6 × 10⁹ years ago: life appearing on Earth as evidenced by vestiges in very old carbonaceous rocks.

About 40 to 30 thousand years ago: the first unequivocal signs of human activity on Earth, remnants of old villages, cultivated land, cattle, burial places, ceramics, calendars, pictorial representations, etc, in various places at Mesopotamia, Egypt, Southern Europe, India and China. From this evidence it can be inferred that men posses intelligence, will and freedom incomparably far above anything in the animal kingdom.

Recorded history shows that men of *all races* and *all cultures* have been conscious that life, health, intelligence, freedom, are *received*. The order discernible in nature has lead them to recognize the existence of a Creator, Life-giver and Law-giver. *Atheism* is not a post-scientific development, it is a consequence of man's freedom to accept or reject that

Creator. Anthony Flew, one of the most honest and respected atheists of the 20th century, held for many years that God cannot in principle be verified or falsified. But, following logic wherever it lead him, late in life, it did lead him to God. About 1985 he began to realize that science was leading him to the same place at which Aquinas had been eight centuries earlier. The classic arguments, according to him, do not offer proof positive for the Christian God. But they strongly suggest that the most conspicuous minds, from Aristotle to Einstein, were on solid ground when they repudiated atheism. (See, for instance, Carl Sundell, "Atheism Yesterday and Today", New Oxford Review, April 2012, pp. 28–32).

The distinguished Spanish American biologist Francisco José Ayala, disciple of Theodosius Dobzhansky (one of the founders of the Synthetic Theory of Evolution) holds that defenders of Creation start with the false idea that the human being is perfect to begin with, and that, taking into account that more than 20 million spontaneous abortions take place in the world yearly, this fact would imply a defective Creator, responsible therefore for a defective reproductive system in human females. The same reasoning would blame the Creator for the death of the majority of human beings at an age between one and one hundred years. But this reasoning is incorrect: the main defenders of Creation, including Catholic, Protestant and Orthodox, hold that God created *freely* and that the created world is not the most perfect conceivable world. God did create everything *in time*, and the four known physical interactions (electromagnetic, nuclear weak, nuclear strong, and gravitational) as well as the set of actual universal constants, imply a cosmic clock which is signalling now that 13.7×10^9 years after the Big Bang, within a finite cosmos, the Milky Way, the Solar System and the planet Earth (including the Biosphere), are evolving continuously at a rate set forth by the fixed magnitude of the physical interactions and by the magnitude of the fundamental constants. But God is free to create or not. And to create one way or another way. To pretend that God is forced to create as he did is tantamount to espouse pantheism.

Every death of a human being is in a way *a tragedy*. But, if human beings (at variance with all animals and plants) have a *soul*, a God-given soul, and have free-will, their death here on Earth may be just the gate to eternal life in Heaven. Evolutionary science, in principle, has nothing to do with it. Free will does.

The theories of Darwin, Marx and Freud as it is well known have pervaded all of modern thought. These theories contain some partial truths, probably much less than half truths, as Chesterton noted about one hundred years ago.

They are also (in many ways) narrow, materialistic, fatalistic and anti-Christian cultural constructs. Darwin's ideas have contributed to a blind belief in "automatic progress", and to foster an all encompassing "survival–of-the-fittest" mentality in social, commercial and political relations. They have been used to discourage charitable support for the less privileged members of human society, and, in the early 20th century, (but not only then) to justify racist policies. Sometimes Darwin has been quoted in this connection as saying that the difference between an Anglo-Saxon white specimen and a pigmy of Africa are greater than between a pigmy and a monkey.

Marx's ideas, as is well known, have produced one hundred million victims in Mao's China, Stalin's Soviet Union and other countries of half the world. And in the other half they have fostered the growth of the authority of the state in detriment of the authority of the family. In affirming that the State is the absolute authority, Marxism has supported state-sponsored compulsory education, replacing religion, family and individual conscience, with disastrous consequences.

Freud's ideas, on the other hand have overemphasized sex out of all proportion, resulting in a serious general decline of morality. They have contributed also to blurr the demarcation between normal and abnormal behaviour. Freud's Psychoanalysis has lead to an incredible rise of the counselling industry of psychotherapy in Europe and America. But serious professionals in the US have noted that the mental health of the average citizen is much worst at present than it was sixty years ago before the tremendous rise on that counselling industry.

As Chesterton notes, if you take away completely the supernatural what remains is often the unnatural. Prophetically, he added:

"Above all, they (Darwinism, Marxism and Freudism) darken it (the mind). All these tremendous and temporary discoveries have had the singular fascination that they were not merely degrading, but were also depressing".

Finally regarding Niels Bohr (Copenhagen) interpretation of Quantum Mechanics, let us quote Henry F. Schaefer, a distinguished contemporary American quantum chemist:

"Bohr implied that there is a sense in which the act of observing creates reality... My view is that this stream of logic is flawed from beginning to end. Molecules emit and absorb photons whether any one observes them or not. Atmospheric chemistry took place long before the techniques of modern chemical kinetics were developed. Molecules existed in interstellar space billions of years before the appearance of mankind, etc, etc, etc.

("Science and Christianity: Coherence of Conflict?"
The University of Georgia, 2003, p.117)

The Editors